Elisabeth Crawford's new study departs from the commonly held notion that universalism and internationalism are inherent features of science. Showing how the rise of scientific organizations around the turn of the century centered on national scientific enterprises, Crawford argues that scientific activities of the late nineteenth century were an integral part of the emergence of the nation-state in Europe. Internationalism in science, both theoretical and practical, began to hold sway over scientists only when economic relations and transportation and communication facilities began to cross national boundaries.

The founding of the Nobel prize in 1901 confirmed the internationalization of science. The workings of the Nobel institution rested on an international community of scientists who forwarded candidates for the prizes. Along with the candidates and eventual prizewinners, they constituted the Nobel population, which, in the fields of chemistry and physics between 1901 and 1939, numbered more than a thousand scientists of greater and lesser renown from 25 countries.

Crawford uses the Nobel population for prosopographic studies that shed new light on national and international science between 1901 and 1939. Her four studies examine critically the following problems: the upsurge of nationalism among scientists of warring nations during and after World War I and its consequences for internationalism in science, the existence of a scientific center and periphery in Central Europe, the effective use of the Nobel prizes in an organization whose primary purpose was to further national science, and the elite conception of science in the United States and its role in the success of the national scientific enterprise. Two introductory chapters provide necessary background by discussing research methodology, and national and international science between 1880 and 1914.

Nationalism and internationalism in science, 1880–1939

Nationalism and internationalism in science, 1880–1939

Four studies of the Nobel population

ELISABETH CRAWFORD

Centre National de la Recherche Scientifique
Group d'Etudes et de Recherches sur la Science, Paris

CAMBRIDGE
UNIVERSITY PRESS

CAMBRIDGE
UNIVERSITY PRESS

University Printing House, Cambridge CB2 8BS, United Kingdom

Cambridge University Press is part of the University of Cambridge.

It furthers the University's mission by disseminating knowledge in the pursuit of education, learning and research at the highest international levels of excellence.

www.cambridge.org
Information on this title: www.cambridge.org/9780521403863

© Cambridge University Press 1992

First published 1992
First paperback edition 2002

A catalogue record for this publication is available from the British Library

Library of Congress Cataloguing in Publication data
Crawford, Elisabeth T.
Nationalism and internationalism in science, 1880–1939: four studies of the Nobel population / Elisabeth Crawford.
p.   cm.
Includes bibliographical references and index.
ISBN 0 521 40386 3 (hardback)
1. Science – Historiography.   2. Science – International cooperation – History – 19th century.   3. Science – International cooperation – History – 20th century.   4. Nobel prizes – Historiography.   5. Scientists – United States – Historiography.   6. Scientists – Europe – Historiography.   7. Competition, International Historiography.   I. Title.
Q126.9.C73   1992
509–dc20   91-33702   CIP

ISBN 978-0-521-40386-3 Hardback
ISBN 978-0-521-52474-2 Paperback

For my son Alexander

Contents

〜〜〜〜〜〜〜〜〜〜〜〜〜〜〜〜〜〜〜〜〜〜〜〜〜〜〜〜〜〜〜〜〜〜〜〜〜

Tables and figures

~~~~~~

Tables

Tables and figures

Figures

Acknowledgments

The idea of using the Nobel population – the approximately one thousand individuals who acted as nominators and nominees for the prizes in physics and chemistry between 1901 and 1939 – for studies in the social history of science arose from solicitations for papers presented at scientific meetings. Two of these papers have been included in collective volumes. I am grateful to the American Institute of Physics for permission to publish materials from my article "Scientific elite revisited: American candidates for the Nobel prizes in physics and chemistry, 1901–1938," which appeared in Stanley Goldberg and Roger H. Stuewer, eds., *The Michelson era in American science, 1870–1930* (New York: American Institute of Physics, 1988). I am equally grateful to the Deutsche Verlags-anstalt GmbH and the Max-Planck-Gesellschaft zur Förderung der Wissenschaften, which hold the copyright to Rudolf Vierhaus and Bernhard vom Brocke, eds., *Forschung im Spannungsfeld von Politik und Gesellschaft: Geschichte und Struktur der Kaiser-Wilhelm-/Max-Planck-Gesellschaft* (Stuttgart, 1990). This book contains an early version of the study of the Kaiser-Wilhelm Society and the Nobel institution by J. L. Heilbron and me. I also want to thank Sage Publications for permission to reprint materials from my article "Internationalism in science as a casualty of the First World War: relations between German and Allied scientists as reflected in nominations for the Nobel prizes in physics and chemistry," *Social Science Information* 27 (1988): 163–201.

The database comprising the Nobel population between 1901 and 1939 was created by the Office for History of Science and Technology (OHST), University of California, Berkeley. I am grateful to the OHST for making resources available to create and manage the database. As

Acknowledgments

the chief "base keeper," Rebecca Ullrich is thanked for her prompt and cheerful handling of my many requests for data.

In many instances, the raw data had to be supplemented with detailed biographical information. I would like to thank the following persons who helped with the hard-to-get information concerning nominators and nominees from east-central Europe: Lubos Novy of the Archives of the Czechoslovak Academy of Sciences, Prague; Gabor Pallo of the Technical University, Budapest; and Alois Kernbauer of the University Archives, Graz. Valuable research assistance was provided by Mahmoud Zamani. Henning Eckart and Marion Kazemi of the Max-Planck Society Archives, Berlin, and Bernhard vom Brocke of the University of Marburg helped with information about the Kaiser-Wilhelm Society, its institutes, and their personnel.

The individuals to whom I want to give special thanks are an anonymous referee for Cambridge University Press who made useful suggestions concerning the draft manuscript; Albert Biderman, who taught me the critical and qualitative approaches to sociology that I have tried to apply in the four studies of the Nobel population; David Cahan, who suggested important revisions of the draft manuscript; and John Heilbron, who graciously let me use the study of the Kaiser-Wilhelm Society, originally published under our joint names, and read and criticized the manuscript.

Introduction

The limited attention given nationalism and internationalism in the history of science makes them the poor relatives of this discipline. The relative neglect of two phenomena that were coincidental with the creation of modern science organization, and also shaped it in so many ways, is not easy to explain. The presupposition that science is and has always been universal – an assumption that will be examined presently[1] – has made inquiries into the influence of nationalism seem irrelevant, even inappropriate. It is somewhat ironic, then, that the most common form of inquiry into the modern science organization that emerged in the late nineteenth century is the national disciplinary history.[2] Also, that despite the universalist ethic with which George Sarton imbued the discipline of history of science, when it was founded early in this century, its practitioners are still billed as historians of French, German, American, or Scandinavian science. However rich in description and detail, their national disciplinary histories are bound to time and place; they are very rarely comparative. On the whole, national science or nationalism in science – and I will show later how the two are related – as an overreaching concept has hardly begun to be explored.

The inquiry into internationalism in science, too, has suffered from the universalist presupposition. The focal point here has been not so much how universalist ideals in science found practical expression in international scientific activities during the latter part of the nineteenth century but the damage done to those ideals during and after World War I. The scientists' not remaining aloof but going to bat for their

[1] See Chapter 2.
[2] For representative examples of national disciplinary histories, see Chapter 1, note 18, and the Bibliographical Essay.

I

nations in the nonshooting, propaganda war, and the breakdown in international scientific relations that followed the war and continued after, have both been seen as tests of how genuine and true-blue the universalist spirit in science really was. The adverse effect of the war is beyond question; in fact, as indicated by the first of the critical and empirical studies in Part II of this volume, the hurt was probably deeper and more lasting than once thought. What is unfortunate is the over-concentration on the war in the historiography of nationalism and internationalism in science.

The budding scholar of nationalism and internationalism in science is not likely to be better served by more general works in political history, social history, or international relations. The overwhelming majority of the authors of these works disregard the sciences, and nearly totally so in the voluminous literature on nationalism.[3] Again, this may be because of the universalist presupposition. A more likely explanation is that scientists constituted a distinct elite within their respective societies and they were therefore few in number. Their persons and activities had little to do with nationalist movements, which appealed mainly to the disfranchised and disgruntled, and among whose members, or programs, historians of nationalism have hoped to find the causes of World War I or II.

This book is an attempt to draw attention to the phenomena of nationalism and internationalism in the history of science, to define each for the purposes of empirical study, and to establish some equilibrium by bringing nationalism into focus and formulating the problem of internationalism in more general terms than the scientists' fall from grace in World War I. These tasks, I believe, can be accomplished only by juxtaposing nationalism and internationalism, trying to analyze them simultaneously. To treat nationalism and internationalism in science from 1880 to 1939 comprehensively is a monumental task, clearly beyond the capacity of the history of science because the empirical materials are not at hand. This volume can only prepare for such a treatise. It tries to do so in three ways: first, by a review of the methods and concepts, in general, those related to the study of scientific development, on which it would have to be based (Chapter 1); second, by providing an overview of nationalism and internationalism in science during the crucial period,

---

[3] An overview of the literature on nationalism is found in the Bibliographical Essay.

Introduction

1880–1914 (Chapter 2); and third by a critique of existing work broadly related to the themes of the book. (The critique is contained in the four studies of the Nobel population that make up Part II.) The inquiry here differs from the customary approach to the Nobel prizes in that it is concerned neither with the selection of prizewinners nor with the significance of the prizes in shaping the public image of science.[4]

The juxtaposition of nationalism and internationalism calls for comparative studies, not just across national boundaries but also on the national as well as on the international plane. The Nobel population, which comprises the approximately one thousand individuals who acted as nominators and nominees for the prizes in physics and chemistry (1901–1939), admirably fulfills these conditions. As both Chapters 5 and 6 demonstrate, the subpopulations of personnel from the Kaiser-Wilhelm Society and American nominators and nominees were part of national elites, if not ultra-elites. At the same time, many of them were active in international networks in their respective fields and thus were highly visible internationally.

The Nobel population is a new data source for the history of science. It came into being when the documents in the Nobel Archives of the Royal Swedish Academy of Sciences (prizes in physics and chemistry) were made available to scholars for purposes of historical research. This occurred in 1974 when the Nobel Foundation relaxed the secrecy provision in its statutes, which applied to the four prize-awarding institutions,[5] and permitted access to archival materials, provided that the documents were at least 50 years old. Because the archives contain only the barest information (name and place of residence), access was only the first step in ascertaining the Nobel population. To serve as a database, biographical and other information had to be added. In 1987, nominal

---

[4] Cf. Elisabeth Crawford, *The beginnings of the Nobel institution: The science prizes 1901 – 1915* (Cambridge and Paris, 1984); Elisabeth Crawford and Robert Friedman, "The prizes in physics and chemistry in the context of Swedish science," in Carl Gustaf Bernhard, Elisabeth Crawford, and Per Sörbom, eds., *Science, technology and society in the time of Alfred Nobel* (Oxford, 1982), pp. 311–331, and Abraham Pais, "How Einstein got the Nobel prize," in Abraham Pais, *"Subtle is the Lord..." The science and life of Albert Einstein* (Oxford and New York, 1982). For other examples, see Robert Marc Friedman, "Text, context, and quicksand: Method and understanding for studying the Nobel science prizes," *Historical Studies in the Physical Sciences* 20 (1989): 63–77.

[5] The four Nobel prizes and the prize in economic sciences in memory of Alfred Nobel are awarded by four institutions: the Royal Swedish Academy of Sciences (physics, chemistry, and economics), the Nobel Assembly at the Karolinska Institute (physiology or medicine), the Swedish Academy (literature), and the Nobel Committee of the Norwegian Parliament (peace).

Introduction

lists of the population were published for the first time.[6] Building up the database with biographical and other information is an ongoing task, inasmuch as the population expands naturally as new archival materials are made available (at present, up to and including 1941). Hence, each year a new vintage of nominators and nominees can be sampled.

The Nobel population of 1901 to 1939, from which the different subpopulations featured in the four studies of Part II are drawn, totals about 950 individuals representing 25 countries. It transcends national boundaries because its members responded to relatively uniform criteria. These were twofold: (1) There were the *candidates*, who were well-known, often eminent, scientists, proposed by their colleagues for their contribution to knowledge; and (2) the *nominators*, who were divided between those with permanent nominating rights and those specifically invited each year to suggest candidates.[7] The ratio of nominators to candidates was about three to one. There was some overlap because the same individual might figure both as nominator and nominee.

The Nobel population is broadly representative of academic physics and chemistry, both research and teaching, during the first three decades of the twentieth century. To appreciate this, one has to consider the small size of the physical sciences enterprise internationally. The number of physicists active in Europe and North America early in the twentieth century has been estimated at one thousand;[8] by the mid–1930s, it may have grown threefold or fourfold. Chemists may initially have been around three times that number and probably grew faster. Between one-fourth and one-third probably figured at one time or another either as candidates or as nominators for the physics and/or chemistry prizes and, hence, entered the Nobel population.

Physicists and chemists active in academe, in particular, university professors, constitute the majority of the population, as do those who hail from the four big science-producing countries: England, Germany, France, and the United States. The population also comprises the per-

---

[6] Elisabeth Crawford, J. L. Heilbron, and Rebecca Ullrich, *The Nobel population, 1901–1937: A census of the nominators and nominees for the prizes in physics and chemistry* (Berkeley and Uppsala, 1987).

[7] For details about the nominating system, see ibid., pp. 1–2.

[8] Paul Forman, J. L. Heilbron, and Spencer Weart, "Physics circa 1900: Personnel, funding and productivity of the academic establishments," *Historical Studies in the Physical Sciences* 5 (1975), whole issue.

sonnel of independent research establishments, such as the Kaiser Wilhelm Institutes in Germany; scientists in government research bureaus, such as the U.S. Bureau of Standards; and, starting in the interwar period, those working in industrial research laboratories.

In each of the four studies a different part of the Nobel population has been examined in detail in order to elucidate a particular problem using the prosopographic method (see Chapter 1). World-historical, geographic, and institutional criteria have been used to select the particular parts of the population examined.

The world-historical population comprises some 950 physicists and chemists from the groupings of countries referred to in World War I as the Allied, neutral, and Central Powers (see Chapter 3). The analysis of their nominations for the prizes in physics and chemistry (1901–1933) sheds light on the effect of the war on the exchange of scientific honors and on internationalism in science more generally.

Geographically based populations are Nobel prize nominators and nominees from east-central Europe and the United States. They are used to examine critically both center-periphery relations within Central Europe and more generally (see Chapter 4) and the claim that the scientific ultra-elite in the United States is coterminous with that nation's Nobel prizewinners (see Chapter 6).

The institutionally based population is represented by the German nominators and nominees associated in various capacities with the Kaiser-Wilhelm Society for the Advancement of Science (Kaiser-Wilhelm-Gesellschaft zur Förderung der Wissenschaften, or KWG) from its founding in 1911 to the outbreak of World War II. Their roles in and for the KWG reveal the particular breed of elite science promoted by the KWG (see Chapter 5).

In general, these studies support the contention advanced earlier that it is only by juxtaposing national and international science that we can hope to take the full measure of both, individually as well as their interactions. The added value is perhaps most manifest in the studies that are primarily nationally based. That no scientific institution, however national or even nationalistic in design and purpose, can function in isolation from international trends is brought home by the analysis of the interworkings of the Kaiser-Wilhelm Society and the Nobel institution. The advantage that the KWG, its members and institutes, sought and found in Stockholm was the enhanced value – international certification, legitimacy, and, perhaps most important, prestige – that the Nobel prizes brought to its activities.

How the international side of science impinges on the national one is evinced even more clearly in the study of scientific elites in the United States. Here, one investigation, that of Harriet Zuckerman,[9] which uses the Nobel prize as the supreme criterion for admission to the American ultra-elite of scientists, is pitted against another, my own, which shows that there are more similarities than differences between the laureates and the nonwinning candidates when one examines the elite attributes brought to the fore by Zuckerman and me. It is significant, though, that irrespective of the discrepancies in our analyses and opinions, we both draw on international reputational measures to define a national elite.

When focusing primarily on the international side of science, again we are well advised to include the national viewpoint. In the study of the international system of nominations for the Nobel prizes in physics and chemistry (see Chapter 3), the upsurge in nationalism observed in the nominations of Allied and Central Power scientists at the time of World War I might have been taken as an instance of chauvinism that would pass once the war was over. However, doing this would have been to overlook the fact that the nominating system was based on the premise that the nominators would act primarily as representatives of their national scientific communities and that a high level of nominations in favor of the nominators' own compatriots was endemic to the system. The national element comes even more directly into play when we try to understand the mechanisms that have made for relatively uniform scientific developments across national boundaries. According to the most widely adopted theory, the explanation is found in the competition and mutual emulation between national scientific communities that have led to the division of the "world of science" into a center (or centers) and a periphery (see Chapters 1 and 4).

In the analysis of scientific development, abstract notions of the inherently universalist character of science have often been coupled with approaches that de facto treat the sciences as primarily national enterprises. It is clear that as analytic stances, both are inadequate. The demands of national science will always be counterbalanced by those of

---

[9] Harriet Zuckerman, *Scientific elite: Nobel laureates in the United States* (New York, 1977).

6

international science, and vice versa. We can begin to understand why one or the other comes to the fore by examining them as different forms of the social institution of science, keeping in mind that the form that scientific activities will take on when organized nationally (in disciplines or specialties, for instance) is not necessarily the same that they will present internationally.

The form that is the main focus here is the reward system of science, in particular, the Nobel prizes, which represent the apex of the hierarchy of honorific awards (prizes, medals, election to scientific societies, and so on). As an important creation of the turn-of-the-century movement toward internationalism, both in science and more generally, the prizes and the Nobel institution are particularly well suited for analyzing interactions between national and international elements in science. These form patterns of a particular kind, but, as in a kaleidoscope, this is only one of an infinite variety of images. It is hoped that the particular kaleidoscopic image produced by these analyses of interactions within and around the international reward system of science will induce alternative images, reflecting other parts of the social institution of science and other times.

PART I

Conceptual and historiographical issues

Methods for a social history of scientific development

In the early 1830s, Alphonse de Candolle, a Geneva naturalist, eager to discover the contribution of different nations to the development of the sciences, invented a new method in the history of science. To gauge the standing of each national scientific group, he counted the foreigners elected to membership in the three major scientific societies – the Paris Academy of Sciences, the Royal Society of London, and the Berlin Academy of Sciences – and calculated the share of the total membership held by each national group. Dividing this share by the size of the country, counted in million inhabitants, gave him a statistical measure of the "scientific value," as he called it, of one million inhabitants in a given country. Although Candolle's measure was necessarily crude and approximate, it was a methodological innovation and a precursor to present-day scientometrics. The four studies of the Nobel population presented in Part II follow a line of inquiry that goes back to Candolle's comparative method and the concepts on which it was based. Broadly speaking, they relate to the social history of science, a field that until recently was not of major interest to historians of science, who were more concerned with ideas and discourse or the "great men" of science.

Not just with respect to methodology, but conceptually as well, Candolle's *Histoire des sciences*... was a precursor in the social history of science when it was finally published in 1873 – for three reasons at least. First, it introduced the notion that sociocultural conditions govern the development of science; second, it related specific national characteristics (religion, class structure, language, type of government, library and other facilities for intellectual work, and public understanding of science) to the degree of "scientific value" of the population of a given country; and third, it examined the prevalence of those characteristics among a preselected group of scientists (those holding foreign membership in the three prominent academies of science in the years 1750, 1789, 1829, and 1869) using the method of collective biography, or prosopography. By and large, Candolle proposed an approach to the history of science that was quantitative, hypothetico-deductive, and comparative.[1]

The blemishes of the work are as notable as its ambitions and achievements. Like many of his contemporaries, Candolle held opinions that in current parlance would qualify as white supremacist, if not racist (he stated, for instance, that "those belonging to the Asian, African and native American races have remained completely outside the scientific movement"); male chauvinist ("no person of the feminine sex has produced an original scientific work"); and Eurocentric ("the non-European races are not to be considered from the scientific point of view").[2] However offensive, these opinions set the initial limitations for the inquiry, but they did not interfere with its execution.

Much of the durable value of the work stems from Candolle's being conscious of the conundrum that he himself and others studying national differences in scientific development face. How could such differences be reconciled, he asked, with the precept that science is universal and has nothing to do with nationalities? To speak of "German chemistry" and "British chemistry" or "French astronomy" and "Italian astronomy," he said, would be to deny the existence of the international republic of science, more real than that of letters because the language and mores of each country mean so little in science. Candolle's solution to his dilemma was ingenious. He argued that the mere fact that he could apply the same concepts and methods when studying scientists in different countries was a reaffirmation of the universal character of science. It constituted proof that science was but little influenced by political or military order, both inherently *national* characteristics.[3]

Candolle made his methodological innovation because the task of surveying the wealth of scientific knowledge that had accumulated in

---

[1] Alphonse de Candolle, *Histoire des sciences et des savants depuis deux siècles* (Paris, 1987), reprint of second edition (Geneva, 1885). See also Xavier Polanco, review of Candolle, *Histoire des sciences, Isis* 80 (1989): 502–503. For the origins and history of the method of prosopography, see Lewis Pyenson, " 'Who the guys were': Prosopography in the history of science," *History of Science* 15 (1977): 155–188.

[2] Candolle, *Histoire des sciences*, pp. 187, 71–72, and 158.

[3] Ibid., pp. 159–160.

his own century and the previous one filled him with despair. By relying on the judgment of the three most prestigious scientific bodies concerning the worth of the scientists selected for foreign membership, he would be freed, he thought, from the burden of assimilating and evaluating this mass of knowledge. The idea of drawing on scientists' judgments of their peers is, of course, at the heart of present-day scientometrics. Where Candolle counted a few hundred members of three national academies by hand, powerful computers now make it possible to map the evaluative structure of entire disciplines or fields, nationally or internationally, using citation patterns.

As Candolle knew, though, by itself, the quantitative method does not suffice to reach an understanding of national differences in scientific development. After having listed the elements that could explain "scientific value" across national boundaries, he therefore described in detail how each element (religion, class structure, language, type of government, and so on) manifested itself concretely in specific countries. In doing so, he used, in crude form, some of the concepts that figure in the studies of the Nobel population. These concepts will receive attention presently.

The problems of studying scientific development

For all its merits, like many pioneering works, Candolle's was beset by problems, primarily conceptual ones. The solutions proposed to some of those problems by Candolle's successors are of interest for what they show of the evolution of both the sciences and the social history of science. I will make no effort to survey the voluminous literature on this subject but will restrict myself to those aspects that have a bearing on the studies presented in Part II.

In Candolle's work, there is no definition of "scientific development." The term is left, instead, as an "empty box," to be filled by whatever empirical materials his inquiry threw up. "Scientific development" was the way the sciences developed; for Candolle, this became differential national developments. The circularity of this logic would no doubt have barred further progress had the ambition been to develop general theories of scientific change. Instead, the social history of science has most often concerned middle-range theory. On this level, statements apply to parts rather than the whole of the sciences, and the temptation toward generalizations is tempered by the requirement that the statements be incorporated in propositions that permit empirical testing.[4] The ambition to go beyond the single phenomenon, be it a great scientist, a laboratory, or a discipline, is one of the characteristics that distinguishes social history of science from mainline history of science.

To the extent that such theories have had an overriding theme, it has concerned the development of the sciences into an autonomous activity practiced by trained professionals. When Candolle launched his prosopographical study in the early 1830s, he could only lament that there was no word in either French, English, or German to designate practitioners of science. He reviewed the existing vocabulary – *savant, Gelehrte, learned men* – and found all the terms too inclusive. He thought it amusing that the absence of a noun to designate "the learned" had forced the English to resort to the French term and to speak, for instance, of "a great savant."[5] He may not have been aware that the term scientist had been invented in 1833 to designate collectively participants in the third meeting of the British Association for the Advancement of Science.[6] It was not until the end of the nineteenth century, however, that the term, and equivalent ones in other languages, gained currency.

The introduction of this term, and corresponding ones in other languages, coincided with the emergence of national scientific enterprises in most countries making up the "civilized world," that is, Europe and North America. When fully developed, these enterprises would encompass all the scientific disciplines, the different kinds of institutions (universities and *technische Hochschulen*) for teaching and research, as well as the various scientific societies for exchanging and publicizing the knowledge produced. Specific dates for the completion of the enterprises are difficult to give; in Germany, Great Britain, and France, it was well before the World War I; in the United States, perhaps only afterward. A major characteristic of a fully developed enterprise was the professionalization of the activity; it became the privileged domain of specialists, whose possession of esoteric knowledge set them apart from other social groups.[7]

---

[4] Robert K. Merton, *Social theory and social structure,* rev. ed. (New York, 1957), p. 9.
[5] Candolle, *Histoire des sciences,* pp. 18–19.
[6] Sidney Ross, "Scientist: The story of a word," *Annals of Science* 18 (1962): 65–85.
[7] See, for instance, Steven Shapin and Arnold Thackray, "Prosopography as a research tool in history of science: The British scientific community 1700–1900," *History of Science* 12 (1974): 1–28. The authors state: "A speculation as to the dating of the ultimate divorce between natural

Methods

Somewhat paradoxically, as the sciences progressed toward isolation from society, the social history of science developed approaches that emphasized the interdependence of science and society. Possession of the tools (concepts and methods) to make generalizable statements was not sufficient; social historians of science also felt the need to understand the meaning of scientific development, in short, to practice the *Verstehen* that Max Weber advocated for historical sociology at the turn of the century. The categories for Candolle's collective biography – and this is another conceptual problem in his work – only allowed him to sort members of his different national populations by social class, religion, education, and scientific tradition in the family. These categories did not go very far toward an understanding of what was specific about scientists individually or as members of a social group, that is, acting both as scientists and members of society.

To arrive at such an understanding required concepts that joined scientific developments and social or societal ones or at least made it possible to analyze them simultaneously. Joseph Ben-David's *The scientist's role in society* has been the most ambitious attempt to join the two.[8] It uses the sociological concept of "role" to examine the social conditions of scientific activity in major science-producing countries over four centuries. "Nationalism" and "internationalism" are other examples of concepts that permit developments in society and in the sciences to be analyzed simultaneously. Such conscious attempts to articulate the concepts used to select and interpret empirical evidence have been rare among social historians of science. Scientific activity is often described in sociological terms – the prime example is the frequent use of the social-organizational and cultural entity known as the "discipline" – but rarely in such a way that the social units studied can be compared or matched with analogous ones found in other social spheres. Instead, these units will appear as unique historical creations, and the statements made about them will not go beyond the single case at hand.

That those analyzing scientific development are now being more ex-

---

knowledge and general culture in Britain would seem to place it between 1870 and 1900" (p. 11). See also Ronald C. Tobey, *The American ideology of national science, 1919–1930* (Pittsburgh, 1971), pp. 31–33.
[8] Joseph Ben-David defines "role" as "the pattern of behaviors, sentiments, and motives conceived by people as a unit of social interaction with a distinct function of its own and considered as appropriate in given situations." Joseph Ben-David, *The scientist's role in society: A comparative study* (Englewood Cliffs, N.J., 1971), pp. 16–17.

plicit about using concepts is probably due to the accrued interest in historical-sociological research generally[9] and to the incursions made into the history of science by sociologists of scientific knowledge. The conceptual, or more often metaphorical, armor of the latter is aimed at the cognitive content of science and its social determinants. For a long time, those studying scientific development left content aside altogether or, like Candolle, resorted to the expedient of relying on scientists' own judgements. Those who confronted the issue either left it to the "scientific method" to determine the social structure of science, as did Robert Merton when he formulated his norms of science, or followed the example of Ben-David and kept carefully separate the social and cognitive spheres of science.[10]

To show that scientific development is socially determined, the sociologists of scientific knowledge often have had to subsume cognitive processes into social ones and to deny the specificity of science. This is illustrated by the use of such concepts as interest, power, authority, and negotiation to understand the motives and mechanisms whereby social actors and situations influence the course of science.[11] As a riposte, a few historians of science have devised concepts that would give *some* social content to the cognitive processes that they consider central to scientific development. The notion of the investigative enterprise, for instance, developed by William Coleman and Frederic L. Holmes, encompasses the facilities, equipment, and resources that have gone into the production of scientific knowledge, as well as the theoretical assumptions, instrumentation, and experimental procedure.[12]

Irrespective of whether they give precedence to social or cognitive elements, historians and sociologists of science agree that more comparative analyses of scientific development are desirable. The problem is how to conduct such analyses. The form of comparative analysis

[9] Victoria E. Bonnell, "The uses of theory, concepts and comparison in historical sociology," *Comparative Studies in Society and History* 22 (1980): 156–173.

[10] Robert K. Merton, *The sociology of science: Theoretical and empirical investigations* (Chicago, 1973), pp. 270–273; and Ben-David, *Scientist's role*, pp. 12, 185.

[11] Programmatic and critical views, respectively, on the use of these concepts are in Steven Shapin, "History of science and its sociological reconstructions," *History of Science* 20 (1982): 157–211; and François-André Isambert, "Un 'programme fort' en sociologie de la science?" *Revue française de sociologie* 26 (1985): 485–508.

[12] William Coleman and Frederic Lawrence Holmes, eds., *The investigative enterprise: Experimental physiology in 19th century medicine* (Berkeley, 1988); and Frederic Lawrence Holmes, *Eighteenth century chemistry as an investigative enterprise* (Berkeley, 1989), p. 126.

initiated by Candolle has been the classical one.[13] His comparisons were made between equivalent units (countries) and involved a range of independent variables (religion, class structure, language, type of government, and so on) that he used to explain differential national developments of science. That the questions he put, and partly answered, are still pressing ones is shown by the agenda that Lewis Pyenson presented to historians of national disciplinary development. "We want to know," he wrote, "why a certain scientific discipline flourished in one setting and not in another." For example, "Why astronomy rose to such prominence in the Netherlands, despite a climate entirely unpropitious for star-gazing, while no such tradition emerged in a topographically well suited country of comparable size and temperament, Switzerland."[14]

Though such comparisons are badly needed, they cannot account for the peculiar features of science that come from its being at the same time both national and international. More precisely, comparisons of nationally based disciplines leave out the autonomous elements of international science – communication networks or supranational research facilities, for instance – that are, of course, more prominent now than they were early in the century. But even then they constituted international superstructures that impinged selectively on national developments.[15]

For all his efforts to grasp the dynamics of national and international interactions on a grand scale, Ben-David's model did not account for these international structures. This is because he viewed scientific development primarily "as a process of diffusion and transplantation of models from *one country to another.*" In this way, his analysis emphasized "the succession of centers [England during the latter part of the seventeenth century, France during the eighteenth century, Germany during the nineteenth century, and the United States at present], instead of systematically comparing the state of science in all countries."[16]

Although put in relation to the overriding concepts of center and periphery, Ben-David's units of analysis were still individual countries,

---

[13] Bonnell, "The uses of theory, concepts and comparison in historical sociology," pp. 164–165.

[14] Lewis Pyenson, "What is the good of history of science?" *History of Science* 27 (1989): 353–389, quote on 379.

[15] Elisabeth Crawford, "The universe of international science, 1880–1939," in Tore Frängsmyr, ed., *Solomon's house revisited: The organization and institutionalization of science.* Nobel Symposium 75 (Canton, Mass., 1990), pp. 251–269.

[16] Ben-David, *Scientist's role*, pp. 16–20, quote on 19 (emphasis added).

whose scientific development was studied in terms of the level of support. By "support," he meant the value placed on science by society and the adequacy of scientific organizations and systems of research. It should be possible, though, to mobilize concepts that compare national and international interactions, not just on an intercountry level but on an intracountry one as well. Three such concepts with particular relevance to the studies presented here are now discussed.

Comparative concepts

The three concepts, and these three are by no means all, that can be used to analyze scientific development both across national boundaries and within countries are the organization of science into disciplines, specialties, and research schools; the elite status of the men and women who practice science; and the division of the world of science into center (or centers) and periphery.

If the concepts used in the social sciences generally are most often bound in time, this is even more true for history, where the concepts have to fit the particular historical problem or epoch studied. A notion such as the religious beliefs of scientists that Candolle and others employed to explain how Protestant cultures came to be in the scientific vanguard, although perfectly valid for the seventeenth and eighteenth centuries, is hardly applicable in the secularized societies of nineteenth- and twentieth-century Europe and America.[17] When the concept no longer fits, it may legitimately be given another meaning. The center-periphery duality, for instance, that is applied in Chapter 4, admittedly by stretching it somewhat, to Europe, in particular east-central Europe, of the early twentieth century, seems valid at present mainly for scientific relations between the industrialized countries and the developing world.

The concepts used here are time-bound in a dual sense: They exhibit the close fit with the period studied (referred to earlier) and they are grist for the mill of the social history of science in the past 25 years. For this second reason, the studies should be regarded as critiques of the service that these concepts have rendered in the past, rather than, as happens all too often in the history and sociology of science, as

---

[17] Candolle, *Histoire des sciences*, pp. 121–127; and Robert K. Merton, "Science, technology and society in seventeenth century England," *Osiris* 4:2 (1938), whole issue.

programmatic statements with respect to the future. In the following, I briefly trace the antecedents of each concept and show its relevance for the comparative study of scientific development.

Disciplines, specialties, and research schools

The most basic, durable, and generalizable construct for analyzing scientific development is the *discipline*. As a basic unit of social and cognitive organization in the sciences, in its modern form it goes back to the mid-nineteenth century; furthermore, its geographic spread is strongly linked to the creation of national scientific enterprises in Europe and America. The story of how those enterprises developed their autonomy is often one of the emergence of disciplines, viewed conceptually, socially, and politically. Pride, power, and national prejudice have been seen as important elements in the drive toward autonomy, as illustrated by the wording of some titles of national disciplinary histories – "the physicists" (in America), "the now mighty theoretical physicists" (in Germany), and "the kaiser's chemists."[18]

Although disciplines as an organizational form can be applied across national and temporal boundaries, this same generalizability limits their ability to explain scientific development. Even such an ambitious attempt at cross-national, disciplinary mapping as "Physics circa 1900"[19] does not go very far toward an understanding of why physics grew (in terms of personnel or funding) or developed specific orientations (experimental rather than theoretical, for instance) in some countries or settings and not in others – that is, the very types of questions raised by Pyenson and by others.

It is possible to gain more insight into these matters by bringing two or more disciplines into play in a cross-national setting. If the similarities of disciplinary development – the "large picture," as it were – are put on the same plane, divergences will emerge more starkly and the causes

---

[18] Daniel J. Kevles, *The physicists: The history of a scientific community in modern America* (New York, 1978); Christa Jungnickel and Russell McCormmach, *Intellectual mastery of nature. Theoretical physics from Ohm to Einstein* (2 vols.), vol. 1, *The torch of mathematics, 1800–1870*, and vol. 2, *The now mighty theoretical physicists, 1870–1925* (Chicago, 1986); Jeffrey A. Johnson, *The kaiser's chemists: Science and modernization in imperial Germany* (Chapel Hill, 1990).

[19] Paul Forman, J. L. Heilbron, and Spencer Weart, "Physics circa 1900: Personnel, funding and productivity of the academic establishments," *Historical Studies in the Physical Sciences* 5 (1975), whole issue.

of these, whether broadly social, cognitive, or other, may be pinpointed. This strategy has been followed in the studies of the Nobel population, all of which involve physics as well as chemistry. To take only one example, in the study of the development of physics and chemistry in east-central Europe (Chapter 4), the similarities in institutional arrangements for teaching and research in both physics and chemistry, which, furthermore, existed for the German "center" as well as the Austro-Hungarian "periphery," made it possible to "freeze," so to say, the organizational background and observe the modes of knowledge production that were specific to each discipline on the periphery.

A more critical stance toward disciplines as the exclusive sites of analysis emerged some 15 years ago among historians and sociologists of science. Two notions, *specialties* and *research schools*, have been proposed as complements, if not alternatives, to disciplines. In discussing the merits of each in explaining scientific change, Gerald Geison pointed to shortcomings that seem particularly pertinent to an analysis of cross-national developments, although such an analysis was not his direct concern. To Geison, focusing on specialties, and especially emerging specialities, revealed "the persistent sociological tendency to focus on new or emerging social units that have no firm geographical locus."[20] He attributed the sociologists' disregard for studying small, localized groups of scientists to Merton's influence, particularly to Merton's contention that, throughout the ages, creative scientists have been "cosmopolitans" oriented toward national and transnational scientific communities rather than "locals" content with their immediate group of associates. Merton urged sociologists of science to develop studies of "larger aggregates of interacting scientists and of spatially distant reference groups and individuals."[21] And so they did. But much of their work became the routine mapping of communication within specialities using sophisticated bibliometric techniques, an exercise that did not, in Geison's opinion "suffice to account for scientific innovation and change."[22] Geison's panacea was to focus on "research schools," not only localized ones but also those that are spatially dispersed, because

---

[20] Gerald Geison, "Scientific change, emerging specialties, and research schools," *History of Science* 19 (1981): 20–40, quote on 31.
[21] Merton, *Sociology of science*, pp. 374–376.
[22] Geison, "Scientific change," p. 20.

to him, "research schools have become the predominant concrete organizational form in science since the mid-nineteenth century."[23]

The choice of schools as the unit of analysis should direct the researchers' attention to more specific problems, and empirical materials, than those of disciplines or specialties. It is surprising, then, to find that the problems Geison formulated for the historians of research schools bore a strong resemblance to the ones Pyenson presented to the historians of disciplines. For example: "Why and how did Liebig's pioneering research school come to be established in so unlikely a setting as the small 'backwater' University of Giessen, while Ira Remsen's school of chemistry was to be a relative failure in so propitious an environment as the new Johns Hopkins University in Baltimore? ... [What] were the ways and means by which Enrico Fermi's group achieved such stunning – if short lived – success in economically depressed and Fascist Italy?"[24]

Disciplines? Specialties? Schools? When the problems studied are those of scientific development or scientific change, viewed more generally in a cross-national perspective, all of these constructs probably should be fielded simultaneously. The necessity for such eclecticism becomes obvious when one considers the multitiered system and the multiple attachments that stem from scientific activities being at the same time national and international. Even our three constructs may not be sufficient to capture the special character of the collaborative channels for international activities. The work of mapping these historically, which has hardly begun, may well call for a fourth category: networks and other forms of informal and transitory groupings.

Elites

Studying scientific development using the concept of *elites* has a longer and more checkered history than any other approach in the social history of science. When late nineteenth-century collective biographers – the most notorious being Francis Galton because of his explicit racism – inquired into the characteristics that set scientists apart as

[23] Ibid., p. 37.
[24] Ibid., p. 27.

an elite, they focused on those that defined scientific genius. To Galton and many others who were inspired by social Darwinism and eugenics, this was more a question of nature, heredity, than of nurture, or environment.

Although the early prosopographers, most of whom are now forgotten, did much to advance the method, not much understanding was gained of the elite characteristics of scientists qua scientists. The populations studied constituted "black boxes," because in whichever way they were selected – drawn from national biographies or from member registers of scientific societies – they were predefined geniuses. This was so because the comparison was not among different groups of scientists – trying to answer the question, for instance, of what made some more successful than others – but between the group of eminent scientists and the population at large. Because much of the work was undertaken to confirm the presumption of a relationship between intelligence and race or class, the greatest emphasis was placed on individual traits, often physiological ones (skin color, brain weight, or longevity, for instance).[25] Candolle was a noteworthy exception, as his measure of the "scientific value" of a national population, and the explanatory variables he used, were structural and sociological ones. To him, the conditions that made for scientific genius were largely social and environmental, not individual. In contrast with Galton, he was not biased toward heredity. After having examined his population with respect to scientific discipline and social origin, he concluded, in fact, that only in mathematics did heredity seem to play a role, and then not even a very important one.[26]

Elite studies were resumed in the late 1960s, this time in the incipient specialty of the sociology of science, more specifically as part of the Columbia Program in the Sociology of Science, directed by Robert Merton. Although prosopography was still the method, the earlier work with its racist overtones had been shelved, this for obvious reasons. Rather than studying scientists as an elite within the population at large, the focus was the scientific elite, that is, the small group of scientists, often Nobel prizewinners, who sat at the top of the pyramid that was the scientific enterprise, in this case the American one. Elite studies,

---

[25] For a review of this largely forgotten literature, see Pyenson, "'Who the guys were,'" pp. 158–162.
[26] Candolle, *Histoire des sciences*, pp. 9–10 and 90–91.

then, became studies of the hierarchical organization of the autonomous scientific enterprise. This hierarchical organization was seen, furthermore, as functional to the progress of science. Because of the importance of the Nobel prize in defining the scientific elite, it seemed fitting that this approach should be the subject of one of the studies of the Nobel population (Chapter 6). For the critique of this approach, the reader is invited to consult this study.

The center-periphery dichotomy

Candolle should be credited with having "discovered," as it were, that scientific achievement was not evenly distributed among the nations that constituted the "civilized world." When he analyzed the geographic provenance of his population of academicians, he found that the majority hailed from countries in the center of Europe (France, Italy, Switzerland, Holland) and not from the periphery (Spain, Portugal, southern Italy, Russia, America). This, he observed, was in contrast to earlier times when illustrious scientists were found in the outer reaches of Europe as well as in the center: Copernicus in Poland, Galileo in Pisa, and Kepler in Germany. He attributed the creation of these clusters of scientific excellence to the nineteenth-century growth of science. It seems, he said, that "the more the sciences progress, the more difficult it is for peripheral or newly civilized countries to do battle with the ones at the center."[27]

Candolle's insight reemerged almost a hundred years later as the grand theme of Ben-David's study. For Ben-David, scientific progress over four hundred years could be subsumed under the succession of countries that had acted as centers for world science: England during the latter part of the seventeenth century, France during the eighteenth century, Germany during the nineteenth century, and the United States since World War II. To Ben-David, the countries that had become scientific centers in modern times were those where the organizational structures for research were built on competition. This produced the innovations that raised the level of scientific activity, not just in the country that had taken the lead, but generally as well. There could be no center(s), of course, without a periphery. To Ben-David, the pe-

[27] Ibid., pp. 246–247.

riphery comprised mainly smaller countries, which tried to copy the organization and orientation of scientific work at the center. But because they lacked a competitive framework for conducting research, they remained at a lower level of scientific performance, just as the center would always be at the top.[28]

Ben-David's model has much to commend it. First, it introduced a comparative approach to scientific development, the dynamic features of which – competition and successive centers, for instance – were built into the model. Second, it forced investigators to specify the vague term "scientific development." They could do so by using Ben-David's notion of the value placed on science by different societies or, as has more often been the case, by mustering quantitative measures of national scientific output, be they discoveries, share of world production of scientific publications, or the perennial counting of the Nobel prizes earned by different countries.[29] Third, and most important, historians of science generally agreed on the existence of centers and on Ben-David's periodization of the shifts from one center to another.

The shortcomings of the model are easy to discern and show up in the work that has been inspired by it. First (and this shortcoming has already been mentioned in the discussion of comparative approaches), the unit of analysis in the model was the nation(s). This induced authors either just to make impressionistic statements[30] or, when empirically based, limit these to the kinds of nationwide indicators of scientific output just mentioned. Second, the model did not address the problems of major scientific powers that are no longer centers. Would they merely continue to live on past glory, or would they actively try to recapture their positions, having more possibilities to do so than countries at the periphery? Third, and probably most important, the model leaves unanswered many questions concerning what is and what is not the scientific periphery.

Although Ben-David made no reference to the third world, studies of center-periphery relations in the past 20 years have largely come to concern those between highly industrialized and developing countries.

---

[28] Ben-David, *Scientist's role*, pp. 172–173.

[29] These measures are found in the appendix in ibid., pp. 186–199.

[30] See, for example, Rainald von Gizycki, "Center and periphery in the international scientific community: Germany, France and Great Britain in the 19th century," *Minerva* 11 (1973): 474–494.

Methods

In this line of inquiry, peripherality has come to mean inferiority and dependence. The periphery is also problematic when viewed historically. Much scientific activity outside Europe and North America in the nineteenth and early twentieth centuries is best described as colonial science and cultural imperialism, demonstrating once more how the sociologists' generalized and abstract concepts have to give way to those that better fit concrete historical situations. The richness, variety, and distinctiveness of science in China over the ages makes it wholly inappropriate to view it in relation to any center-periphery polarity. It also remains an open question if the term periphery is applicable to countries like those of Scandinavia that have a long tradition of scientific development, even if, in the late nineteenth and early twentieth centuries, they were intellectually close to, and to some extent modeled themselves on, the German center.[31]

The study of center-periphery relations in Central Europe (Chapter 4) was done in a critical spirit, giving the center-periphery dichotomy a second chance, as it were, by testing it empirically. The level of analysis was not that of countries, then, but of disciplines (physics and chemistry) within countries and across national boundaries. Scientific development was observed and measured along three dimensions: institutional arrangements for teaching and research; communication of research results and the recognition this brought through citations; and orientation of work. Because the institutional development and, hence, the settings, mainly university ones, for teaching and research were similar throughout Central Europe, dissimilarities along the other two dimensions appeared more starkly.[32] In order not to anticipate the conclusions of the study, it will only be said here that a considerably nuanced view resulted, as rather than the polarity between the center and the periphery postulated by Ben-David, the relations between the two were found complementary in many respects.

---

[31] For contrasting views on the peripherality of Swedish science, see Robert Marc Friedman, "The Nobel prizes and the invigoration of Swedish science: Some considerations," in Tore Frängsmyr, ed., *Solomon's house revisited*, pp. 193–207, and Elisabeth Crawford, *The beginnings of the Nobel institution: The science prizes 1901–1915* (Cambridge and Paris, 1984), pp. 32–37.

[32] The technique of comparison by focusing on differences has been advocated by Marc Bloch and his followers. Marc Bloch, *Land and work in medieval Europe, Selected papers by Marc Bloch* (Berkeley/Los Angeles, 1967), p. 58. See also Lewis Pyenson, "Pure learning and political economy: Science and European expansion in the age of imperialism," in R. P. W. Visser, H. J. M. Bos, L. C. Palm, and H. A. M. Snelders, *New trends in the history of science*. Proceedings of a conference held at the University of Utrecht (Amsterdam and Atlanta, 1989), pp. 212–213.

Conceptual and historiographical issues

Prosopography applied to the Nobel population

The method of prosopography, or collective biography, that Candolle introduced into the history of science is also the one used in the four studies of the Nobel population presented in Part II. Prosopography is a method to describe the biographical features of a population quantitatively and present them in tabulations. The purpose may be merely to draw a numerical profile of the entire population by showing its preponderant traits, or, if one is more ambitious, to cross-tabulate those traits or to analyze them with respect to subgroups within the population. The particular route followed, as well as the complexity of the analysis, depends, of course, on the research problem at hand. On the whole, historians of science, not all of whom are convinced of the value of the method, have practiced a simple and straightforward form of prosopography, one that does not involve sophisticated statistical measures.

To think that one writes history, though, merely by doing prosopography – or, as Walther Struwe put it: "I have set out to explore how biography *becomes* social and intellectual history"[33] – is, of course, a dangerous delusion. No such transsubstantiation is likely to occur, and prosopography is, therefore, best used in conjunction with other methods. The most common one is the historical narrative that places the population studied in its context – institutional, intellectual, or political – and analyzes it in relation to this context. Other examples are historical periodizations, for instance, the chronology of events marking the flux and reflux of internationalism in science between 1900 and 1933 presented in Chapter 3 or the fuller biographical data on individual scientists sprinkled throughout this book. But other quantitative methods can also be used to enrich prosopography. In the history and sociology of science, the most "natural" complement is citation analysis used, for example, in Chapter 4.

The biographical and other data mustered in the studies represent conventional prosopography; for example, they chart education, training, and career trajectories; give typologies of research orientations; and map affiliations with research schools or circles. Dates of birth serve to construct and compare age cohorts. Two kinds of intrapopulation compar-

---

[33] Walther Struwe, *Elites against democracy: Leadership ideals in bourgeois political thought in Germany, 1890–1933* (Princeton, N.J., 1973), p. 9 (emphasis added).

26

ison are common to all the studies. One is the comparison between Nobel prizewinners and unsuccessful candidates and nominators. The other, the sorting of individuals or their nominations by nationality, uses the convention adopted for the Nobel population as a whole; that is, it refers to the country where the person was working when he or she entered the population. Persons are considered as working in a country other than their own if they have spent a minimum of eight years in that country, and their new nationality is assigned retroactively from the time of arrival in the country. In most cases, the person was working in the country of which he or she was also a national.[34]

The periods covered in the four studies fall between 1900 and 1933–1939, in particular, 1914 to 1933. Because most members of the Nobel population were born around the mid-nineteenth century or even earlier, their biographies provide data that have made it possible in two of the empirical studies (Chapters 4 and 5) to go back to the 1880s. However, this does not do full justice to the period between 1880 and 1914, which was crucial in the development of modern culture and science. This was the time when national scientific enterprises were constituted and with them the different forms of nationalism in science. It was also the time when the internationalist ethos and international practice took hold in science. In the period between 1880 and 1914, one can see, then, how the two themes of this book, nationalism and internationalism, emerge both as friends and foes. It is to this subject that we turn in Chapter 2.

[34] Elisabeth Crawford, J. L. Heilbron, and Rebecca Ullrich, *The Nobel population, 1901–1937: A census of the nominators and nominees for the prizes in physics and chemistry* (Berkeley and Uppsala, 1987).

2

∾∾∾

First the nation: national and international science, 1880–1914

The general terms nationalism and internationalism carry so many meanings that it is necessary to try to specify these before applying the terms to the sciences. Of the two, the terminology surrounding nationalism is the more perplexing; in the words of one author, it is "a phenomenon so Protean in its manifestations as to defy precise definition."[1] In common language, the term is most often taken to mean the waving of flags to solidify sentiments of national allegiance and to mobilize against foreigners. This form of nationalism is most closely associated with the period 1880 to 1914, not only because it was then that many nationalist movements came into being and spread throughout Europe, but also because this was the great time of nationalist doctrine. As both the movement and the doctrine gained currency, they bred the more virulent form of nationalism that came to be known as chauvinism.[2] It is in this flag-waving, rabble-rousing sense that nationalism in the sciences has most often been described – and decried.[3] The term can be given another sense, though, one that does not refer to doctrine or movement, but to nationalism as a phenomenon inherent in certain sociopolitical conditions. As defined by Ernest Gellner, who has propounded this alternative view: "Na-

[1] Michael Hughes, *Nationalism and society: Germany, 1800–1945* (London, 1988), p. 3. For overviews of the voluminous literature on nationalism, see the Bibliographical Essay.
[2] E. J. Hobsbawm, *The age of empire, 1875–1914* (London, 1989), pp. 142–164. The Frenchman Chauvin, a grenadier in Napoleon's Grande Armée, whose devotion to the emperor was praised in propaganda pamphlets, lent his name to chauvinism. (Jan Romein, *The watershed of two eras: Europe in 1900* [Middletown, Conn., 1978], p. 106.)
[3] Brigitte Schroeder-Gudehus, "Nationalism and internationalism," in G. N. Cantor et al., eds., *Companion to the history of modern science* (London and New York, 1989), pp. 909–919.

tionalism is primarily a political principle, which holds that the political and the national unit should be congruent."[4] This definition seems to fit best the development of national science enterprises in the late nineteenth century and the types of nationalism among scientists that accompanied them. It will receive more attention presently.

If nationalist movements and doctrines sprang to life in the late nineteenth century, so did internationalism, although its constituency was the well bred and well educated rather than the disfranchised. The term internationalism came to stand for the ideological belief that governments and peoples could act constructively together, particularly in order to abolish war and conflict, and the profusion of practical international activities undertaken, with more or less success, in almost every field of social activity.[5]

In a similar manner, internationalism in science partakes of both theory and practice; as such, it is unproblematic. The confusion arises from the way the term scientific internationalism is used interchangeably, or mixed up, with universalism. The universalist ethos holds that the acceptance or rejection of knowledge claims is totally independent of the personal attributes – sex, race, nationality, religion, or social class – of those who make them. Universalism has been seen as specific to the sciences because it is rooted in the "scientific method," in particular, the judging of knowledge claims according to preestablished, impersonal criteria. Because the method makes the sciences inherently universalist, they are, in the opinion of one author, "by nature international,"[6] or, in the fuller, more idealistic version of another author:

> Because of its objectivity, science is considered to be supracultural, untrammelled by the conflicts of values to which all other expressions of culture are dedicated; in the same way, its universality leads to the postulation of a supranational specificity for the institution which it constitutes and for its members. Even the idea of national scientific communities is contradictory; there can only be one scientific community, which must therefore be international. A single language, similar procedures, com-

---

[4] Ernest Gellner, *Nations and nationalism* (London, 1988), p. 1.
[5] F. S. L. Lyons, *Internationalism in Europe, 1815–1914* (Leyden, 1963).
[6] Schroeder-Gudehus, "Nationalism and internationalism," p. 909; and Diana Crane, "Transnational networks in basic science," *International Organization* 33 (1971): 585–601, quote on p. 585.

Conceptual and historiographical issues

parable experiments, shared norms – all these characteristics must distinguish scientific activity from any other.[7]

These views are open to attack on two fronts. First, they are built on the canon of a unique and universally applicable scientific method, which is a questionable presupposition in the opinion of those who claim that science is socially constructed and that conflicting knowledge claims are settled through negotiation.[8] The latter is the most important issue, for if this contention should hold up (which is far from having been established empirically), it would throw doubt on the most crucial test of scientific universalism – that is, in the words of Robert Merton, the chief expounder of the norm, "*sooner or later*, competing claims to validity are settled by universalistic criteria."[9]

Second, these views confuse science as an abstract method for establishing universally valid knowledge, with science as an activity and a social institution. The nature of scientific discourse, especially the striving for reproducibility and generalizability of results, has brought an ease of transfer of the products of science from one language and culture to another.[10] This has favored the ethos of universalism in science more than in other social spheres, but it has not meant that it rules unchallenged over all aspects of science, that is, as method, social activity, and institution, at all times. On the contrary, over the past one hundred years, most professional scientific activities have centered on national institutions, and their financing has often been linked to national purposes.

The analogy with economic activities is instructive. In the late nineteenth century, both capital and economic goods increasingly crossed national borders, speeded by new transport facilities but also by capitalist

[7] Jean-Jacques Salomon, "The 'internationale' of science," *Science Studies* 1 (1971): 23–42, quote on p. 24.
[8] See, for example, Karin Knorr-Cetina and Michael Mulkay, eds., *Science observed: Perspectives on the social study of science* (London and Beverly Hills, 1983), pp. 11–12, 142, 150–151; Bruno Latour, *Science in action* (Milton Keynes, 1987), chapter 3; and Karin D. Knorr, Roger Krohn, and Richard Whitley, *The social process of scientific investigation*. Sociology of the sciences yearbook, vol. 4, 1980 (Dordrecht, 1981).
[9] Robert K. Merton, *The sociology of science: theoretical and empirical investigations* (Chicago, 1973), 271n (emphasis added).
[10] Lewis Pyenson, "Pure learning and political economy: Science and European expansion in the age of imperialism," in R. P. W. Visser, H. J. M. Bos, L. C. Palm, and H. A. M. Snelders; *New trends in the history of science*. Proceedings of a conference held at the University of Utrecht (Amsterdam and Atlanta, 1989), pp. 210–211.

theories, which held that an international division of labor created the most favorable climate for economic growth. There was an evolving world economy, which did not preclude, however, that national economies also thrived. As in science, these were the main arenas for the production and consumption of goods and for financing. As will be shown presently, one observes a similar coextension of national and international science.

First the nation

The definition of nationalism propounded by Gellner suggests an approach to nationalism in science that does not view it as an aberration, but as a historical phenomenon linked to a certain stage of sociopolitical and economic development. Its contours appear most clearly in Europe and North America in the high industrial age of the late nineteenth century. Not coincidentally, this was also the period when the foundations of modern science organization were laid.

The pivotal point of Gellner's definition is the congruence of the nation and the state giving rise to that peculiar nineteenth-century construction, the nation-state. Arriving at such a congruence is the ambition of all nationalist movements, unless they are permanently relegated to the diaspora. But the achievement of statehood is not the end of nationalism. The view of nationalism as an ongoing process is illustrated by the declaration of Jozef Pilsudski when he assumed leadership of newly independent Poland in 1918: "It is the state that makes the nation and not the nation the state."[11]

This process of "creating" nations is at the heart of Gellner's formulation. It is a process that is by no means restricted to the political sphere. Culture and language are the most essential elements, but not in the sense of the folk cultures and dialects encouraged by romantic nationalist movements. On the contrary, in Gellner's words:

> Nationalism is essentially the general imposition of a high culture on society, where previously low cultures had taken up the lives of the majority, and in some cases the totality, of the population. It means that generalized diffusion of a school-mediated, academy-supervised idiom, codified for the requirements of reasonably precise bureaucratic and tech-

---

[11] Quoted in Hobsbawm, *Age of empire*. p. 148.

nological communication. It is the establishment of an anonymous, impersonal society, with mutually substitutable atomized individuals, held together above all by a shared culture of this kind, in place of a previous complex structure of local groups.[12]

To Gellner, it was industrialization that created the new division of labor and the large social units that made necessary the shared high culture, which defines a "nation." In the industrial age, he concludes, "such a nation/culture becomes the natural social unit...and cannot normally survive without its own political shell, the state."[13]

Gellner's model is attractive because it points to the many different functions that science assumed in the service of the nation-states of Europe and North America from the 1860s onward. On the practical level, there were those linked to industrialization: Physics contributed theoretical knowledge, measures and standards to the all-important electrical industry; chemistry provided new processes and products in the dyestuff, food, metallurgy, and explosives industries; the physical sciences in general were sources of knowledge, instrumentation, and, especially, trained workers in transportation, mechanics, communications, and trade. No less important than their work for the nation's material welfare was the role of science and scientists in creating the "high culture" that made for the unity of the nation-state: upholding the tradition of the nation's great universities while at the same time making those venerable institutions a force of modernization; being *Kulturträger* (bearers of culture), infusing scientific knowledge and values into the cultural life of the nation; and generating the discoveries that would make the citizenry in general identify with and be proud of its scientists.

Germany would be the strategic site, of course, the "laboratory," as it were, to test how well Gellner's model fits the development of a national science. Most of the materials to do so exist. German science and scientific institutions in the Wilhelmian era are well documented, and the so-called German problem has produced an abundance of historical work concerning the forms of nationalism that preceded and, in the opinions of some historians, forebode national socialism.[14] Such an inquiry (which can only be sketched out in barest outline here) would

[12] Gellner, *Nations and nationalism*, p. 57.
[13] Ibid., pp. 142–143.
[14] For the relevant literature, see the section on Germany in the Bibliographical Essay.

focus on the industrial growth and cultural nationalism that were dominant in Germany from the 1870s to World War I. Unification and the proclamation of the *Kaiserreich* in 1871 had only created the political shell that was the state; this now had to be filled with the material well-being and cultural coherence that would make for a nation. This latter carried the dual complication that the Reich included peoples (the Poles of East Prussia, for instance) who were not part of the German "nation," defined in linguistic terms, and excluded those who were (the Austrians).

The sciences participated in creating material well-being in the ways described earlier and many others as well: The new science-based industries were important factors in economic growth and in the industrialization that was in full swing only after 1871. France was outdistanced in industrial production in 1880 and England in 1900. In 1913, Germany produced as much industrial goods as the two taken together.[15]

First and foremost, though, the contributions of the sciences to creating the shared high culture that would make for national identity and cohesiveness involved the university, not only its new functions in mass education (shown by a sixfold increase in the number of science students between 1871 and 1912) but also defending the ideals on which it was based, in particular, those of *Lehrfreiheit* and *Lernfreiheit*.[16] The universities, and especially the science faculties, were the chief exponents of the success of German science brought about by combining teaching and research. The handsome laboratories and institutes – 23 were constructed in physics alone between 1872 and 1915 – that grew up adjacent to the main university buildings were the material proof.[17] They attracted foreign students, around 8 percent of the total number, which was a larger contingent than in any other national university system. Not surprisingly, this caused the envy of French university scientists.[18] That the

---

[15] Bernhard vom Brocke, "Die Kaiser-Wilhelm-Gesellschaft im Kaiserreich: Vorgeschichte, Gründung und Entwicklung bis zum Ausbruch des ersten Weltkriegs," in Rudolf Vierhaus and Bernhard vom Brocke, eds., *Forschung im Spannungsfeld von Politik und Gesellschaft: Geschichte und Struktur der Kaiser-Wilhelm/Max-Planck-Gesellschaft* (Stuttgart, 1990), p. 19.

[16] Ibid., p. 22. *Lehrfreiheit* refers to the right of university professors to engage in teaching and research freely, unrestricted by the requirements for degrees and diplomas; *Lernfreiheit* refers to the right of students to register and study at the university of their choice.

[17] Ibid., and David Cahan, *An institute for an empire: The Physikalisch-Technische Reichanstalt, 1871–1918* (Cambridge, 1989), pp. 20–21.

[18] Harry Paul, *The sorcerer's apprentice: The French scientist's image of German science, 1840–1919* (Gainesville, 1972), pp. 15–20.

mission of the universities and the sciences was nationalist, not only in the cultural sense but in the flag-waving one as well, was demonstrated by the opening of the German university in recently annexed Strasbourg in 1872 (reinaugurated in 1877 as the Kaiser-Wilhelm Universität zu Strassburg) and the funds that were lavished on it by the Reich, especially, for the construction of scientific research institutes and laboratories.[19]

Although in these areas relations with the nation-state were mediated by the universities, in other ones scientists became much more directly involved in bureaucractic and regulatory functions – to take only one example: the standardization of weights and measures, not just the kilo and the meter used in trade and commerce, but also standard electrical units, which were national concerns before they became international ones. The German constitution of 1871 provided for the establishment of a Bureau of Standards (Normal-Eichungs-Kommission). In 1887, the joint interests of physicists, government officials, and industrial leaders in precision measurement and technology (that is, the construction of instruments) led to the creation of the Imperial Institute of Physics and Technology (Physikalisch-Technische Reichsanstalt).[20] The idea that the state should provide for the needs of science, directly and not mediated through the universities, came to fruition slowly, expressed in the new term science policy (*Wissenschaftpolitik*), used for the first time in 1900.[21] It blossomed in the establishment in 1911 of the Kaiser-Wilhelm Society for the Advancement of the Sciences (discussed in Chapter 5).

It is important to keep in mind that much of the initiative that led to the regulatory and bureaucratic functions of science came from the scientists themselves, often the most prominent members of academe. Already in 1862, a decade before the Reich had come into existence, the most prominent of them, the physicist and physiologist Hermann von Helmholtz set forth the tasks of science in support of national power. He stressed its dual nature: residing not only in industry and material welfare but also in political, moral, and cultural values. The state, if it

---

[19] John E. Craig, *Scholarship and nationbuilding: The universities of Strasbourg and Alsatian society, 1870–1939* (Chicago, 1984), pp. 29–67, 73.

[20] Cahan, *An institute for an empire*, chapters 2 and 3, and Peter Lundgreen, "Wissenschaft als öffentliche Dienstleistung: 100 Jahre staatliche Versuchs-, Prüf- und Forschungsanstalten in Deutschland," in Vierhaus and vom Brocke, eds., *Forschung im Spannungsfeld*, pp. 673–691.

[21] vom Brocke, "Die Kaiser-Wilhelm-Gesellschaft," p. 20.

was to prosper, had to support not only the natural sciences and technology but also the social sciences and the humanities. This support was not disinterested, however, because it was also the duty of scientists to work "for the good of the entire nation . . . on its order and at its costs." Helmholtz's claim that scientists formed "a sort of organized army" was prophetic because it pointed to the most important relationship of science and the state in the twentieth century, that of service to the nation at war.[22]

Although nationalism and national science were particularly prevalent in Germany from 1871 to 1914, due to the concurrent developments of industrialization and nation building, they were, of course, not unique to the Reich. The strength of Gellner's model lies in what it suggests of the generalizability of the phenomenon and, therefore, the possibility to observe and compare its multiform manifestations in different national settings at different times. The following very preliminary description of three types of nationalism in science in the late nineteenth century may serve as a guide to future comparative studies.

The first was the *mature*, or to use Gellner's term *natural, nationalism* found in nation-states of such long-standing, relatively speaking, as England, France, the Netherlands, and Switzerland. This was the form of nationalism toward which the new states of Germany and Italy were heading. National scientific enterprises existed with varying degrees of autonomy.[23] A high degree of autonomy meant that scientific practice was insulated from national needs and concerns, at least in peacetime. Whether scientists were patriotic, or even chauvinistic like most of their fellow citizens, or belonged to the minority who did not feel the appeal of chauvinism[24] did not matter much in their work. They could emphasize that their role as patriots in no way interfered with the trans-

---

[22] Hermann von Helmholtz, "Über das Verhältniss der Naturwissenschaften zur Gesammtheit der Wissenchaften: Akademische Festrede gehalten zu Heidelberg beim Antritt des Prorectorats 1862," in Hermann von Helmholtz *Vorträge und Reden*, 4th ed., 2 vols. (Braunschweig, 1896), vol. 1, pp. 159–185. (Quoted in Cahan, *An institute for an empire*, pp. 66–67.) See also David Cahan, "Helmholtz, science, and politics in Germany, 1860–1900," paper presented at the Annual Meeting of the History of Science Society, 1988.

[23] The complex relations between French scientists and the centralized state bureaucracy offers much insight into how autonomy is achieved. See Harry W. Paul, *From knowledge to power: The rise of the science empire in France, 1860–1939* (Cambridge, 1985).

[24] To quote Hobsbawm, this minority was made up of "international socialists, a few intellectuals, cosmopolitan businessmen and the members of the international club of aristocrats and royals" (*Age of empire*, p. 160).

mission of their products across national boundaries, as the Frenchman Louis Pasteur did in his famous saying: *"Le savant a une patrie, la science n'en a pas."* Still, the scientific milieu and products most familiar to them were national ones; furthermore, these were most often restricted to their own disciplines. Evidence of this "myopia" is found, for instance, in the high proportion of nominators (more than 60 percent) who *in peacetime* nominated fellow nationals for the Nobel prizes in physics and chemistry (see Chapter 3).

It should not be forgotten, though, that this mature form of nationalism was practiced in the age of empire, and that all the countries cited above (with the exception of Switzerland) enlisted science and scientists for their imperialist strategies. What these were, how many scientists were enrolled, and with what effects of national scientific enterprises are matters that have been described by others.[25] That these activities are referred to as "cultural imperialism," though, suggests a close kinship with cultural nationalism that would be well worth exploring.

The second form of nationalism in science might be termed *practical nationalism,* to describe a closer involvement of science and scientists with national interests and concerns than in the mature type. Practical nationalism was found outside the European scientific centers, on the northern periphery (Denmark, Norway, and Sweden) and in North America (Canada and the United States). In Sweden, for example, the belated but rapid industrialization that occurred in the 1880s and 1890s accelerated the creation of modern national science. The movement toward modernization of the country set a new agenda of work for the traditional disciplines of natural history. It did so, although to a lesser extent, for physics and chemistry as well. Geology, mineralogy, and geography took on new tasks of surveying, mapping, and inventorying the natural environment. Physics found use in meteorology, and chemistry spawned such new important subfields as agricultural chemistry, wood chemistry, and metallurgy. These tasks were carried out alongside those of building new scientific institutions or of reforming existing ones.

[25] Lewis Pyenson, *Cultural imperialism and exact sciences: German expansion overseas, 1900–1930* (New York, 1985); Lewis Pyenson, *Empire of reason: Exact sciences in Indonesia, 1840–1940* (Leyden, 1989); Pyenson, "Pure learning and political economy," pp. 209–278; Nathan Reingold and Marc Rothenberg, eds., *Scientific colonialism: A cross-cultural comparison* (Washington, D.C., 1987).

First the nation

Because the scientists who took on these tasks were few and far between, their roles were polymorphous; they were at one and the same time researchers, administrators, entrepreneurs, politicians, and *Kulturträger*.[26] What defines their form of nationalism is, first of all, their practical actions, but also the way these involved the nation and national sentiments not only culturally but in a physical-spatial sense. Here, science linked to the management of natural resources was seen as helping to expand the inhabited part of the national territory, making the wilderness cede to "civilization," to use a cliché, whether it be in the American West or the Far North.[27]

The third form was the *militant nationalism* that spread throughout Europe in the late nineteenth century as "nations" without states – Poles in Germany and Russia, Finns in Russia, Czechs and Slovenes in the Austro-Hungarian Empire, Welsh and Irish in the United Kingdom – found themselves at odds with increasingly powerful state bureaucracies.[28] The movements, based on nationalist sentiments, emphasized cultural and linguistic distinctiveness as a means to awaken nationhood and eventually to achieve statehood.[29] Scientists participated in these movements as part of the intelligentsia, most actively when they devoted themselves to creating and maintaining the institutions of higher learning that were seen as important "homes" for the nation. In Wales, for example, the national university, set up in 1893, was the first, and for a while, the only national institution of the Welsh people.[30] The splitting of Charles University and *technische Hochschule* in Prague into a German and a Czech part in 1868 and 1882, respectively, was symptomatic of the way militant nationalism had come to dominate the Habsburg multinational empire.[31] Scientists were involved qua scientists, for instance, in their choice of language of publication.

---

[26] Gunnar Eriksson, *Kartläggarna: Naturvetenskapens tillväxt och tillämpningar i det industriella genombrottets Sverige, 1870–1914*. Acta Universitatis Umensis, 15 (Umeå, 1978); and Elisabeth Crawford, *The beginnings of the Nobel institution: The science prizes 1901–1915* (Cambridge and Paris, 1984), pp. 30–42.

[27] Sverker Sörlin, *Framtidslandet: Debatten om Norrland och naturresurserna under det industriella genombrottet*. Acta Regiae Societatis Skytteanae 33 (Stockholm, 1988), chapter 5.

[28] Hobsbawm, *Age of empire*, pp. 142–164.

[29] Not surprisingly, the important role of the intelligentsia as "awakeners to nationhood" has been stressed by a Czech political scientist. See Miroslav Hroch, *Die Vorkämpfer der nationalen Bewegung bei den kleinen Völkern Europas* (Prague, 1968). See also E. J. Hobsbawm, *Nations and nationalism since 1789: Programme, myth, reality* (Cambridge, 1990), p. 104.

[30] Hobsbawm, *Age of empire*, p. 157.

[31] See Chapter 4.

37

Conceptual and historiographical issues

Going international

In view of the preponderance of national science and nationalism in science, in the sense given the term in our discussion, it is certainly pertinent to ask: Was there any room for internationalism in science? The answer is that, indeed, there was, for in the period of about 1870 to 1914, there was not only a growing belief that the sciences had a special contribution to make to international ideas, but also an upsurge in practical international scientific activity. Much historical writing about international science errs in confounding the two, which is to mistake rhetoric and high aims for action and accomplishment. The latter can be ordered along an axis. At one end, one finds scientists' spontaneous international coordination of their own activities; at the other, activities, often government-induced, where the element of competition was much stronger. This is a schematic representation, of course, for in many activities coordination and competition blended into one.

The largest part of the upsurge of international science activities in the period 1870 to 1914 arose from activities of coordination, in particular, international congresses. This is why they could be considered international, if we define an international activity (arbitrarily, of course) as involving scientists from three nations or more. Had these activities taken the form of the research collaborations, that is scientists working in teams on a common project, which only occurred in the interwar period or even after World War II, they would have been much more likely to be bilateral or trilateral. For a long time, there was only one supranational research facility, the International Bureau of Weights and Measures (Bureau International des Poids et Mesures) at Sèvres outside Paris, established in 1875, which did more testing than research.

International science in the period discussed here was extensive rather than intensive; as such, it was an outgrowth of the diffusion of the products of scientific work through correspondence networks and journals. Its chief centers were, in order of importance, international congresses; the international societies, whose chief function was to organize these; and some common projects, often government-sponsored, undertaken to coordinate and improve observation and measurement.

For international congresses in the natural and physical sciences, the list of "firsts" starts in 1860 with the congress of chemists called by August Kekulé in Karlsruhe to bring some coherence and consistency into chemical notations. It continued with botany and horticulture (1864), geodesy (1864), astronomy (1865), pharmaceutical sciences (1865), meteorology (1873), and geology (1878). In the 1880s and 1890s, hardly a year went by without at least one meeting of scientists in one field or another. These were no longer "firsts" but regularly instituted congresses held every two or three years in major European cities, often in conjunction with international exhibitions. Of the 91 international meetings held in conjunction with the Paris Universal Exhibition of 1889, 15 were scientific ones. Scientists were so prolific in their travel, in fact, that the French-based Association for Scientific Advancement (Alliance Scientifique Universelle), founded in 1876 as an overall co-ordinating body, issued a form of identity card or passport, the "Diplôme-Circulaire," which scientists could carry when they traveled abroad.[32]

Scientists' travel, as well as that of other members of the middle or upper classes to which they belonged, was made possible by the transport revolution, in particular the railway boom. Between 1870 and 1914, the world's railway network expanded from a little more than 200,000 kilometers to more than one million. In the early 1880s, almost two billion people traveled on them. Trains not only took scientists to their congresses, on their visits with colleagues and to laboratories, and on their postdoctoral sojourns in different countries, they also carried their mail – personal letters, journals issued abroad, or manuscripts for publication in these journals – at a speed that should put present-day mail services to shame. Steamship travel across the Atlantic, or to the British or French colonies, lagged somewhat behind the rail travel because, in the 1880s, sailing ships still made up the bulk of the world's shipping tonnage.[33] This changed rapidly, for by the turn of the century scientists were crossing the Atlantic with the same ease that they had been crisscrossing Europe for the past 20 some years. The bevy of European scientific luminaries – Svante Arrhenius, Ludwig Boltzmann, Paul Langevin, Wilhelm Ostwald, Henri Poincaré, and William Ramsay – who descended on St. Louis in 1904 to attend the Congress of the Arts and Sciences,

---

[32] Lyons, *Internationalism in Europe*, pp. 223–224; Anne Rasmussen, "Jalons pour une histoire des congrès internationaux aux 19ème siècle: Régulation scientifique et propagande intellectuelle," *Relations internationales* 62 (1990): 115–133; and Bernadette Bensaude-Vincent, "Karlsruhe septembre 1860: L'atome en congrès," *Relations internationales* 62 (1990): 149–169.

[33] Hobsbawm, *Age of empire*, pp. 27–28, 62.

held in conjunction with the Universal Exhibition, and then fanned out across the American continent were only too typical representatives of the new scientific tourism.

All this served the purpose for which it was intended: to communicate and coordinate scientific work, so that not only new results but also innovations in instrumentation and experimental procedures would spread throughout the scientific "world," that is, Europe and North America. Then, as now, scientists felt that these purposes were best served by face-to-face contact. The most tangible and most useful result was the standardization of nomenclature, methods, and units. This need was first felt in the natural sciences. Although skeptical of international congresses, Candolle, writing in the 1870s found them "necessary to establish unity in matters that are of interest to all countries, for instance, geodetic measures, meteorological measures or how to set the first meridian." To give results, though, there has to be a definite international interest and advance preparation, otherwise the congress will "degenerate into feasting and ceremony."[34]

Toward the end of the nineteenth century, the creation of a transnational laboratory culture resting on the experimental parts of the physical and biological sciences made standardization an urgent matter in these disciplines, too, and also favored its execution.[35] The Karlsruhe congress of chemistry of 1860 was the first of many similar efforts to standardize chemical notations. Standardization also figured importantly at the international congresses in chemistry that began to be held regularly, starting in 1889. As the problems became more complex, they were referred to specialized, permanent committees: the International Committee on Atomic Weights (1897), the International Commission of Photometry (1900), the International Committee for the Publication of Annual Tables of Constants (1909), and the International Radium Standard Commission (1910), to mention only a few.

All these efforts to coordinate and organize internationally contained an element of internationalism. There is no doubt that the creation of

[34] Alphonse de Candolle, *Histoire des sciences et des savants depuis deux siècles* (Paris, 1987), reprint of second edition (Geneva, 1885), p. 153.
[35] William Coleman and Frederic Lawrence Holmes, eds., *The investigative enterprise: Experimental physiology in nineteenth century medicine* (Berkeley, 1988); and Crawford, "The universe of international science, 1880–1939," in Tore Frängsmyr, ed., *Solomon's house revisited: The organization and institutionalization of science.* Proceedings of Nobel Symposium 75 (Canton, Mass., 1990), p. 259.

international scientific organizations was part of a movement that embraced nearly every kind of social activity. This is demonstrated by the covariance of the growth of scientific organizations established each decade between 1855 and 1914 with the overall increase in international organizations during the same period. Both grew at the same rate, the total number of organizations doubling every 10 years from 1875 onward, and that of scientific organizations from 1885 onward. In the last decade before the war close to 25 organizations were created compared to 10 in the previous decade. The "mortality rate" of international scientific organizations was also approximately the same as that of the total number of organizations (60 percent for scientific organizations and 70 percent for all organizations).[36]

To gauge the role of internationalism in practical international science, it would be necessary to carry out detailed studies of specific organizational initiatives and ventures. The rhetoric that surrounded these often masked the significance of both motives and achievements. The International Association of Academies (IAA), for example, was certainly steeped in internationalist rhetoric, yet it accomplished almost nothing from the time it was created in 1899 until it was disbanded during World War I.[37]

Alongside coordination, competition was an important stimulus for international scientific activities, most directly so when national interests were involved. Government involvement did not necessarily mean a strong competitive element, for, as F. S. L. Lyons has pointed out, governments were prepared to give financial assistance to international collaborative projects, provided they "did not cost too much, that the scientists themselves were prepared to do the work, and that nothing in the commitment trenched upon national security or sovereignty."[38]

There were a few such intergovernmental collaborative ventures that yielded practical results before World War I. They had in common that the initiative emanated from a national institution keen on associating foreign groups and researchers with its activities. This was the model for the international project to measure the earth that was initiated by

---

[36] Lyons, *Internationalism in Europe*, pp. 14–16.
[37] Ibid., pp. 225–228; and Brigitte Schroeder-Gudehus, "Division of labor and the common good: The International Association of Academies, 1899–1914," in Carl Gustaf Bernhard, Elisabeth Crawford, and Per Sörbom, eds., *Science, technology and society in the time of Alfred Nobel.* Proceedings of Nobel Symposium 52 (Oxford, 1981), pp. 3–20.
[38] Lyons, *Internationalism in Europe*, pp. 228–229.

the Prussian Institute of Geodesy (Preussisches Institut für Erdmagnetismus) in 1862 with the encouragement of the Prussian government. It eventually led to the creation of the International Geodetic Association (1867), which did not become truly international, however, until 1886. The technical work continued to be centered on the Prussian Institute, whose director was also the director of the permanent bureau of the association, but from 1895 onward, it was financed by contributions from various governments. The International Association of Seismology (1903) is another example of building an international organization and running collaborative projects; its bureau was located at the Imperial Seismological Station (Kaiserlichen Hauptstation für Erdbebenforschung) in Strasbourg and headed by George Gerland, the station's director. A more equitable sharing of duties among participating countries was found in the areas of astronomy, meteorology, oceanography, and solar research, all of which gave rise to international associations and ambitious collaborative projects before World War I.[39]

The competitive element was much stronger and government involvement more direct when scientific activities were perceived as promoting trade and commerce. We saw earlier how important the standardization of weights and measures was for industrialization and how nation-states enlisted physical scientists for that task. In the latter part of the nineteenth century, these national interests became objects of international action. The French government led the way through its campaigns for the metric system, which culminated in the International Congress on Metric Standards of 1875. It produced the Metric Convention, whose signatories contracted to adopt the metric system and, in return, gained seats on the International Committee on Weights and Measures. The latter was set up to promote the metric system and to oversee the research in metrology carried out at the International Bureau of Weights and Measures.[40]

Commerce and trade, and, hence, national rivalry, were even more present in the work to establish international standards for electricity. It took four major international conferences, starting with the one in

[39] Ibid., pp. 229–234.
[40] Charles Edouard Guillaume, "Les systèmes de mesure et l'organisation internationale du système métrique," *La vie internationale* 3 (1913): 5–44; Charles Edouard Guillaume, *La création du Bureau International des Poids et Mesures et son oeuvre.* Ouvrage publié à l'occasion du cinquentenaire de sa fondation (Paris, 1927).

42

Paris in 1881, until the actual standards to be employed for the judiciously chosen triad of ampere, ohm, and volt were adopted at the International Electrical Congress held in Chicago in 1893. The extension of national interests and institutions into the international sphere was manifest in the German delegation, which represented a merging of the triumvirate interests of government, academe, and industry: Wilhelm Förster, director of the Imperial Institute for Weights and Measures; Hermann von Helmholtz, director of the Imperial Institute of Physics and Technology; and Werner von Siemens, founder and owner of the giant of the electrical industry Siemens and Halske.[41]

Internationalism and nationalism
in the context of the Nobel institution

During the period 1880 to 1914, nationalism and internationalism in science coexisted to a degree that had not occurred beforehand and was not to recur. How the two interacted and were reconciled in a specific setting is brought to light in the Nobel institution, which started to function in 1901, the approximate midpoint of the period examined here.

From the time of its creation, the Nobel institution stood for internationalism in science and culture. Alfred Nobel himself put the institution on this course when he specified in his will (1895) "in awarding the prizes no consideration whatever shall be given to the nationality of the candidates, but that the most worthy shall receive the prize, whether he be a Scandinavian or not." That the corporations, which awarded the prizes in science and medicine, were able to live up to this lofty goal can in large part be attributed to the impressive progress made in international scientific communication in the preceding years. (It helped, of course, that there were, indeed, startling results to communicate.) The prize awarders were thus presented on a platter with discoveries and scientists who were already well known internationally: W. C. Röntgen with his X rays, Becquerel and the Curies with radioactivity,

---

[41] Cahan, *Institute for an empire*, pp. 13, 122; Andrew Butrica, Paolo Brenni, Christine Blondel, and Peter Lundgreen, *Standardization and units in electricity, 1850–1914*, papers from an international workshop held on July 6, 1988, Centre de Recherche en Histoire des Sciences et des Techniques, Cite des Sciences et de l'Industrie (Paris and Lancaster, 1989); Christine Blondel, "Les premiers congrès internationaux d' électricité," *Relations internationales* 62 (1990): 171–182.

W. Ramsay and (Lord Rayleigh) J. W. Strutt with the noble gases – to mention only the first years' prizewinners. No wonder, then, that the first years' prize ceremonies became celebrations of scientific internationalism and that the institution as a whole came to be shrouded in internationalist rhetoric and symbolism.

Given the role of internationalism in the life and lore of the institution, it is easy to forget that it was built not only on the coexistence of nationalism and internationalism but also on the essential tension between the two. That science was essentially national was acknowledged in the provision (paragraph 7) in the statutes of the Nobel Foundation (1900) that scientists, acting primarily as representatives of their national scientific communities, be invited to designate candidates for the prizes. Here national considerations and competition between nations could be expected to be important in the choice of nominations, but it could also be hoped that the nominators would follow Nobel's precept and not let their judgment be influenced by nationality.[42]

Somewhat incongruously, in view of the internationalist ambitions for the institution, the selection of prizewinners was left in the hands of local corporations made up of Swedish nationals. In physics and chemistry, the Nobel committees (one for each discipline) evaluated the works proposed for the prizes and recommended the most prizeworthy to the Royal Academy of Sciences. The final decision was placed in the hands of the one hundred Swedish members of the academy meeting in plenary session.[43] Cast in the role of arbiters of the achievements of scientists representing different nations, the Swedish prize awarders saw themselves as the guardians of the internationalist spirit of the institution. That they acquitted themselves of this task with honor was due in large measure to their disregard of the nominating system when it, as will be shown in Chapter 3, became infused with chauvinism during and after World War I. The prize awarders could ignore the system because a provision (paragraph 7) in the statutes of the Nobel Foundation merely specified that to be considered for the prize, an individual had to be proposed in writing by "a person competent to make such nominations." It was thus possible to receive the prize, as did Nils Dalén, the Swedish

---

[42] Crawford, *Beginnings of the Nobel institution*, pp. 101–108.
[43] Ibid., pp. 81–84.

Table 1. *Distribution of Nobel prize candidates, nominations, and winners in physics and chemistry by country, 1901–1933, in percentages* (N *in parentheses*)

Country	Candidates		Nominations		Winners	
Germany	32	(108)	37	(990)	37	(26)
France	16	(53)	20	(538)	16	(11)
Great Britain	13	(42)	10	(264)	20	(14)
United States	11	(38)	11	(285)	7	(5)
Sweden	4	(13)	3	(80)	7	(5)
Netherlands	3	(9)	2	(61)	6	(4)
Italy	3	(9)	3	(76)	0	(0)
Other Nordic countries[a]	6	(19)	6	(146)	1	(1)
Central and Eastern Europe[b]	7	(24)	4	(121)	3	(2)
Other countries[c]	5	(18)	4	(114)	3	(2)

[a]Denmark, Finland, and Norway.
[b]Austria, Czechoslovakia, Hungary, Poland, Russia, and Yugoslavia.
[c]Argentina, Australia, Belgium, Canada, India, Japan, Spain, and Switzerland.

inventor (Ph 1912; here, and following, Ph refers to the Nobel prize in physics, and Ch to that in chemistry) on the basis of only one nomination. The prize juries' autonomy in relation to the nominating system is brought out in the following discussion.

Table 1 gives the distribution of candidates, nominations, and prize-winners among the countries or groups of countries participating in the nominating process. Had the prize awarders selected the laureates on the basis of the "votes" received by candidates from each country, the proportion of laureates from a given country should have been roughly the same as the percentage of nominations received by candidates from that country. Instead, as brought out in Table 1, for all countries and groups of countries except Germany, the percentage of prizewinners does not even roughly equal the percentage of nominations received by all candidates from these countries. That some countries (Great Britain, the Netherlands, and Sweden, for example) received more prizes than they would have had the prizes been assigned on the basis of a plurality of votes, and other countries (France, Italy, and the United States) less, indicates that the prize juries in Stockholm exercised their own judgment. The most important consideration in the evaluation of candidates was their scientific specialties and their contribution to these specialties.[44] The juries were probably also influenced by the notion that the historical list of prizewinners should include representatives of all "civilized" nations.

The comparison of the votes received by the winners and nonwinners shows that the Swedish prize awarders, whether consciously or not, emphasized internationalism by giving preference to candidates who were nominated in large part by other than their own compatriots. Whereas the prizewinners received 83 percent of their nominations from countries other than their own, the nonwinners received only half this percentage from noncompatriots, that is, 42 percent. The 83 percent figure is inflated, as it comprises many smaller countries (Sweden and the Netherlands, for instance) whose winners received important support from the outside. When limited to major powers (Germany, France, Great Britain, and the United States), the corresponding figures are 53 percent for the winners and 40 percent for the nonwinners. That the winners enjoyed stronger international support was due not only to the prize awarders' setting store upon such support but also no doubt to international reputation being important at all stages of the prize selection process. For the prize juries to function as supranational arbiters, however, prizeworthy works had to be brought to their attention, preferably supported by scientists from different countries. This presumed the smooth workings of the international communication networks through which such information had to pass. That these functioned well during the first decade and a half was of paramount importance in establishing the prize's international reputation.[45] World War I caused serious disturbances, however, as will be shown in the first study (Chapter 3) of the Nobel population.

[44] Ibid., pp. 150–187.
[45] Ibid., pp. 203–205.

PART II

Critical and empirical studies

3

Internationalism in science
as a casualty of World War I

In the history of international science, World War I has been regarded as the ultimate test of universalism, a test that in the opinion of most observers the scientists of the belligerent nations failed. That the war should have brought international scientific activities, including the award of the Nobel prizes, to an almost complete halt is not considered surprising, nor is it surprising that scientists served in the armed forces either in research and development or by fighting and dying in the trenches. What is at issue is the way many prominent scientists, particularly in Germany, sought and found their field of action in propaganda warfare, exhibiting the virulent form of nationalism known as chauvinism.

Acting together with scholars in the humanities and with writers and artists, they signed manifestoes, such as the Appeal of the Ninety-three Intellectuals issued in 1914, setting forth German war aims or repudiating Allied charges of German war crimes. Acting alone, they, as well as their Allied counterparts, wrote articles and pamphlets denigrating the claims of superior scientific achievement made by the other side.[1] Many tried to persuade their correspondents, particularly when these

---

[1] The most important manifestos signed by German scientists are reprinted in Klaus Böhme, ed., *Aufrufe und Reden deutscher Professoren im Ersten Weltkrieg* (Stuttgart, 1975). See also Lawrence Badash, "British and American views of the German menace in World War I," *Notes and Records of the Royal Society of London* 34 (1979): 91–121; Bernhard vom Brocke, "Wissenschaft und Militarismus," in William M. Calder III, Hellmut Flashar, and Th. Lindken, eds., *Wilamowitz nach 50 Jahren* (Darmstadt, 1985), pp. 647–719; Harry Paul, *The sorcerer's apprentice: The French scientist's image of German science, 1840–1919* (Gainesville, 1972), pp. 29–86; Brigitte Schroeder-Gudehus, *Deutsche Wissenschaft und internationale Zusammenarbeit, 1914–1928* (Geneva, 1966); Brigitte Schroeder-Gudehus, *Les scientifiques et la paix: La communauté scientifique internationale au cours des années 20* (Montreal, 1978).

were scientists in neutral countries such as the Netherlands and Sweden, that any or all of such actions were justified.

These activities prepared the ground for the postwar politics of international science, which became dominated by the Allied-sponsored International Research Council (IRC) set up in 1919. The policies of ostracism advocated by the IRC reflected a widespread opinion that Germany should be chastized. Hence, in the early postwar period German scholars could or would not attend international meetings or otherwise participate in international scientific activities.

Such a massive intrusion of politics into the supposedly nonpolitical realm of science naturally left scars. Even today, the boycott – the term coined for the series of measures advocated by the IRC – remains a sensitive issue for many scientists, to be evoked primarily as a warning to show what happens when the norms of universality and organized skepticism are set aside. The statement of A. J. Cock, that "scientists should look back with some, possibly prophylactic, shame on the events of 1918 and 1919" is not atypical of such soul-searching.[2] My concerns here do not lie with the moral lessons to be drawn from these events but with the historical record. That this record is still partly in dispute relates mostly to the way the key issues for research into this period have been formulated.

To date, most of the historiography of international scientific relations in the interwar period has concerned the politics of science with the IRC as the chief actor.[3] In some of these studies, the successes and failures of the boycott are seen as the test of whether or not scientific internationalism survived the war.[4] The validity of this test in the narrow sense is not at issue because most historians agree that the prewar organization of international science was a casualty of the war (among many other organizations, it caused the demise of the International Association of Academies). The question is, rather, whether this test is adequate as a gauge of the strength and depths of nationalist feelings that the war had unleashed among scientists of the belligerent nations.

[2] A. J. Cock, "Chauvinism and internationalism in science: The International Research Council, 1919–1926," *Notes and Records of the Royal Society of London* 37 (1983): 249–288.
[3] Ibid.; Schroeder-Gudehus, *Les scientifiques et la paix;* Paul Forman, "Scientific internationalism and the Weimar physicists: The ideology and its manipulation in Germany after World War I," *Isis* 64 (1973): 151–180; Daniel J. Kevles, "'Into hostile political camps': The reorganization of international science in World War I," *Isis* 62 (1970): 47–60.
[4] See, for example, Cock, "Chauvinism and internationalism," pp. 249–250.

Restricting our view to the IRC and its boycott of scientists from the Central Powers means looking mainly at the "general staff" of international science, many of whom were officials at the national academies, and leaving aside the "foot soldiers" in the different scientific fields. Furthermore, in applying the test, historians, lacking other data, have most often based their evaluations on the public statements of scientists. Unfortunately for those who would like to see a clear-cut outcome, for each example of a prominent scientist or scientific organization inveighing against a counterpart in former enemy countries, others can be found that are indicative of much more conciliatory attitudes.

Our study here examines the flux and reflux of internationalism and nationalism, or even chauvinism, among the scientists from the Allied nations, Central Powers, and neutral countries (for explanations of these terms see Figure 3), who nominated candidates for the Nobel prizes in physics and chemistry between 1901 and 1933. Because it uses a large population of scientists (approximately 950 in all) in Europe and North America, who are studied over a period of 30 years, the study departs significantly from the test-case approach that has restricted inquiry to the IRC and the boycott (that is, the period 1919 to 1926).

Research methodology

At the time of the outbreak of World War I, the Nobel prizes were already regarded as the apex of the system (also known as the reward system of science) that honored nationally and internationally prominent scientists with prizes and medals, membership in leading academies of science at home or abroad, honorary doctorates, and invitations to commemorative occasions. The act of nominating candidates for the Nobel prizes, then, was part of the honors that prominent scientists frequently bestowed on each other. The nominations involved a dual distinction: Not only did those who were invited to nominate feel honored, but they, in turn, could honor their colleagues by proposing them for the prizes and also letting them know that they had done so. This was an important function of the nominations, perhaps even more important than their use in the process of prize selection, for (as discussed in Chapter 2) the prize awarders did not consider the nominations as votes, but exercised their own judgment.

Looked at as an exchange of honors, the considerations that made for a nominator's choosing a candidate from a foreign country can be seen in the context of the relations that constituted the fabric of international scientific life, relations that were torn asunder in World War I. In view of how little is known of the specific reasons for nominators' choices, however, there are important limiting conditions to be set on the nominating data.

The statistical analysis presented in the following discussion is based on a classification of each nomination by the nationality of the nominators and nominees.[5] Restricting the analysis to nationality necessarily blocks out a host of other considerations – personal, scientific, and political or most likely an amalgam of the three – that also entered into a nomination. Furthermore, nationality preferences are analyzed statistically through aggregate data that show trends and patterns but do not reveal if individual choices were motivated by nationality and, if so, why. The results of the statistical analysis prove that nationality considerations were important. They also show that these were contingent upon world historical events. This emerges even more clearly when a qualitative interpretation is added to the statistical data drawing on sources (correspondence, biographical writings, pamphlets, manifestoes, and newspaper articles) that reveal how prominent scientists, individually or as a group, took position with respect to nationalism and internationalism in science and politics. From these materials, the specific considerations that made nominators choose, or not choose, in terms of nationality can also occasionally be gleaned.

However, underneath the flux and reflux, largely contingent on external events, there is the phenomenon of a consistently high level of nominations in favor of the nominators' own compatriots ("own-country nominations") and a correspondingly low level of nominations given candidates in other countries. For the French nominators, for instance, own-country nominations remained around 80 percent for the entire period of 1901 to 1930. The French case is instructive because it points up some more permanent considerations behind choices in terms of nationality that were not restricted to the nominators of France. We will

---

[5] As explained in Chapter 1, the nationality designations used for candidates and nominators refer to the country in which they were working and not to nationality per se. A person is considered as working in a country other than his or her own if he or she has spent a minimum of eight years in that country. In most cases the person was working in the country of which he or she was also a national.

touch upon three such considerations: "favorite son," insularity, and reciprocity.

The *"favorite-son"* label can be attached to candidacies that attracted the majority of the votes of the nominators from a given country, usually for a number of years. Small, isolated scientific communities are the likeliest breeding grounds for favorite sons.[6] In France, this kind of candidacy seems to have been more frequent than in any of the other major scientific powers. The best documented case is that of Henri Poincaré, the mathematical physicist, who was the favorite French candidate for the physics prize in the years shortly before his death in 1912. In 1910, for instance, all of the 14 French physicists nominating for the prize put forth Poincaré.[7] This kind of support naturally encroached on the votes that otherwise might have gone to physicists of other nationalities.

The high level of own-country nominations by French nominators may possibly be explained by their inadequate knowledge of the achievements of physicists and chemists in other countries – in other words, their *insularity.* French physics in the interwar period has been described as a *"monde fermé,"* cut off from the international physics community.[8] If a high level of own-country nominations is taken as an indicator of insularity, the French scientific community could be seen as already having turned inward early in the century. In using this indicator, however, it is important to keep in mind that the men and women who figured as nominators were the leaders of French science (Aimé Cotton, Maurice and Louis de Broglie, Frédéric and Irène Joliot-Curie, Paul Langevin, and Jean Perrin, to cite only a few from the interwar period). Their exchanges with foreign scientists whom they visited abroad or met at home meant that they could not have been unaware of the kind of high-level work of scientists who were likely to be put forth for the Nobel prizes. Why, then, did they not suggest foreign candidates? A possible explanation can be found in our third consideration.

A general feature of most sociointellectual and professional relationships, *reciprocity* has been stressed as particularly important for exchanges

---

[6] See Chapter 4.

[7] Elisabeth Crawford, *The beginnings of the Nobel institution: The science prizes 1901–1915* (Cambridge and Paris, 1984), pp. 140–148.

[8] Dominique Pestre, *Physique et physiciens en France, 1918–1940* (Paris and Montreux, 1984), pp. 149–168.

in the French cultural context. "Everything is stuff to be given away and repaid," wrote Marcel Mauss in his classic essay "The Gift."[9] Applying Mauss's ideas to modern French society, Wilton Dillon showed how French industrialists in the 1950s strove to recast both the overall aims of the Marshall Plan and their own role in the plan into exchange relationships. Similarly, Daniel Lerner has taken note of French expectations of reciprocity in interview situations.[10]

Given the scarcity of resources (positions, facilities, and grants) in the French university system during the early part of the century, it is not surprising that honors and recommendations for honors took on importance as commodities not to be given away freely but to be *exchanged.* Furthermore, the web of organizations – so much more interwoven in France than in Germany or England – making up the science establishment provided both donors and recipients with the institutionalized means of fulfilling their obligations toward each other. The prestige of the Nobel prize and the high status of the French scientists who were invited to nominate candidates made nominations for the prize a particularly valuable commodity. Hence, bestowing such a precious gift on foreigners with whom French scientists were not in close enough contact for exchanges to occur (see the argument of insularity) would have been tantamount to throwing it away.

International scientific relations
and Nobel nominations: an overview

To gauge roughly, and by no means exhaustively, the disturbances in international scientific relations caused by World War I, it is useful to conceive of these relations as involving four levels of interaction and to examine representative events on each level.

The first level, mostly governmental, had as a key event the decision taken by the entente after the war to supplant the International Association of Academies, founded in 1899, with the International Research Council, which practiced an exclusionary policy with respect to the

---

[9] Marcel Mauss, "Essai sur le don, forme et raison de l'échange dans les sociétés archaiques," *L'Année sociologique* 1 (new series) (1923–1924): 30–186.
[10] Wilton S. Dillon, *Gifts and nations: The obligation to give, receive and repay* (The Hague and Paris, 1968), pp. 30–186; and Daniel Lerner, "Interviewing Frenchmen," *American Journal of Sociology* 62 (1956): 187–194.

former Central Powers. Although this is the best documented part of international scientific relations, this organizational change did not initially have much effect on scientists' work. Many probably agreed with T. W. Richards, the American chemist, who declared himself in 1919 "not being an enthusiast of the IRC..." and went on to state (in a letter to Svante Arrhenius): "I think that new ideas do not come through any amount of winding of 'red tape' (which, you know, is our slang term for the routine and formality of official direction)."[11]

The second level reflects the results of organizational efforts to bring together scientists from different countries, for example, in order to hold congresses and meetings. The *third level* is private and therefore difficult to document systematically. Here one finds the laboratory visits and letters exchanged across national frontiers, and the swapping of research results, experimental materials, and students. Finally, there is a fourth level, the exchange of honors through Nobel nominations, which is the main focus of this study.

The increase in the number of congresses and meetings constitutes a convenient measure of activities on the second level. As discussed in Chapter 2, this was the primary form of international scientific communication and collaboration in the prewar period. In the last three decades of the nineteenth century, about 20 international scholarly meetings, including scientific ones, were held each year; in the 15 years preceding the outbreak of World War I, this number had risen to 30. During the war, only seven international meetings were held. In the first five years following the war, however, the prewar level was rapidly reestablished and surpassed: By the early 1930s, the mean number of congresses per year had increased by a factor of three.[12] There was an

11 T. W. Richards to Svante Arrhenius, October 8, 1919, Arrhenius Collection, Centrum för Vetenskapshistoria, Kungl. Vetenskapsakademien (Center for History of Science, Royal Academy of Sciences), Stockholm (hereinafter referred to as KVA); Schroeder-Gudehus, *Les scientifiques et la paix*, p. 238; Daniel Kevles, "The International Research Council, 1914–1931: A study in politics, science, and organizational failure," paper given at Conference on Science, Government, and Internationalism, University of California, Berkeley, 1970.
12 Siegfried Grundmann, "Zum Boykott der deutschen Wissenschaft nach dem ersten Weltkrieg," *Wissenschaftliche Zeitschrift der Technischen Universität Dresden* 14 (1965): 799–806; "Deutsche Wissenschaft und Ausland in der Statistik. I. Internationale Wissenschaftliche Kongresse und Organisationen," *Forschungen und Fortschritte* 9 (1933), 330–332. This article gives the number of scholarly meetings held in different countries during a much longer time period (1845–1932) than that covered by Grundmann (1922–1926). There is reasonably good agreement, however, between the figures given in the two articles during the period of overlap. It should be noted that neither author separates out scientific meetings from general scholarly meetings.

important difference, however, in that before the war, international meetings had been open to all qualified and interested persons regardless of nationality, whereas most of the decade following the war was characterized by restrictive practices with respect to German scholars and the organization of meetings in Germany.

It is often taken for granted that the exclusion of German scientists from international meetings was the result of the injunctions issued by the IRC to its member organizations. The most effective boycott of Germany occurred in the immediate postwar years, however, when there were only the three constituent unions of the IRC (astronomy, geodesy and geophysics, and chemistry), which were mostly paper organizations. That the boycott rested, instead, on the general disapprobation of rampant militarism in the German scholarly milieu during the war is shown by the slow return of Germany to international meetings as these sentiments dissipated. Whereas in 1919 Germany was not represented at any international scholarly meetings, by 1922 to 1924, the exclusions applied to 66 and in 1925 to 47 percent of the meetings. A breakthrough occurred in 1926, when the exclusions dropped to about 15 percent of the meetings. Here, again, the actions of the IRC – in this case, its decision to lift the bar against the admission of the former Central Powers – were not the cause but rather the symptom of a normalization of international scientific relations.[13]

The road back to relative normalcy traveled by three individual scientists – Max Planck, Wilhelm Wien, and Paul Langevin – can be used to illustrate the third level, that of private contacts. It also shows the key phases and conditions of the period of reconstruction. Because the role of prominent scientists in repairing relations among the former enemy countries depended in equal measure on opportunity and individual initiative, there were no typical actors on the scene. The three we discuss here, all physicists, resembled each other, though, by the relative importance of their international contacts before the war and their willingness to resume those contacts after the hostilities were over.

Being too old for active duty, Planck and Wien joined the war on the propaganda front. They both signed the mass appeal of 3,016 German

---

[13] Grundmann, "Zum Boykott," pp. 799–806; "Deutsche Wissenschaft und Ausland," pp. 330–332.

academics in support of militarist values and were among the select group of 15 scientists, 7 of whom were Nobel prizewinners, whose names, to their great discredit in the eyes of most Allied and also some neutral scientists, figured on the Appeal to the Cultured Peoples of the World, otherwise known as the Appeal of the Ninety-three Intellectuals.[14] Planck soon modified his nationalist stance, but Wien did not relent. Together with 15 other physicists (among them the incorrigible nationalists Philipp Lenard and Johannes Stark), he issued a manifesto in 1915 which declared that scientific relations with England would not be resumed for the foreseeable future, that German physicists should not publish articles, only rejoinders in British journals, and that they should cite English authors only when absolutely necessary. In 1916, he was again agitating with Lenard, this time against the *scandalosum* of the French language only being used in the invitations to nominate candidates for the Nobel prizes.[15]

The slow defeat of Germany made for repentance, and when conditions permitted it – the armistice having been followed by aborted coups d'état in Berlin and Munich – private visits, facilitated by old personal ties, were exchanged between scientists from the former enemy countries. In 1921, Ernest Rutherford visited Wien in Munich on his way to a summer in the Alps. In 1915, Rutherford had written (to Svante Arrhenius): "It seems to me . . . that all social and scientific intercourse with Germany will be practically stopped for this generation." Similarly, Wien had declared (in a letter to C. W. Oseen): "We will never again know the old relationships."[16] Realizing the hardships that German scientists experienced in the immediate postwar period, Rutherford arranged to have British journals sent to Wien and also gave him a small supply of helium.[17] Planck's first meeting with an Allied scientist after the war probably occurred at a formal occasion, the Nobel prize cere-

---

[14] vom Brocke, "Wissenschaft und Militarismus," pp. 650–664.
[15] J. L. Heilbron, *The dilemmas of an upright man: Max Planck as spokesman for German science* (Berkeley, 1986), pp. 69–81; "Aufforderung," Archive for History of Quantum Physics, Office for History of Science and Technology, University of California, Berkeley; Wilhelm Wien to Philipp Lenard, July 29, 1916, Wien Papers, Deutsches Museum, Munich (hereinafter referred to as DM).
[16] Ernest Rutherford to Svante Arrhenius, June 1, 1915, Arrhenius Collection, KVA; Wilhelm Wien to C. W. Oseen, November 19, 1914, cited in Wilhelm Wien, *Aus dem Leben und Wirken eines Physikers* (Leipzig, 1930), p. 60.
[17] Ernest Rutherford, "Erinnerungen," in Wien, *Aus dem Leben und Wirken*, pp. 152–154.

mony in Stockholm in 1920, which brought together the German scientists who had won prizes during the 1914 to 1919 period and one Allied scientist (C. G. Barkla).

By the mid–1920s, the ground had been tested sufficiently for public semiofficial relations to be reestablished. Wien traveled to England in 1925 to give the tenth Guthrie lecture before the Physical Society of London. He used the occasion to pay a return visit to Rutherford in Cambridge. Later the same year, Planck, Wien, and other German physicists joined with colleagues from the former Allied powers to honor H. A. Lorentz on his jubilee by contributing to the Lorentz Foundation for the Promotion of the Interests of Theoretical Physics. They did so, however, only after having received assurances that there was no danger of protests by any official body of scientists – that is, academies or other member organizations of the IRC – these having been carefully kept out by the Dutch organizers of the fund.[18] The breakthrough in international scientific relations around 1926 is illustrated by Wien's and other German scientists' participation in the annual meeting of the British Association for the Advancement of Science. Consulted on the matter, Planck thought it was right to attend if the invitation applied "not to the individual person, but to German scientists in general." Planck's putting this policy into practice led to the carefully negotiated participation of four of the leading German theoretical physicists in the fifth International Solvay Meeting (1927), which signaled an official return into the international physics discipline.[19]

Some ten years younger than Planck and Wien, Langevin participated in the war on the research and development front where he made a major contribution to submarine detection by inventions in the area of ultrasonics. The war changed him into a pacifist, but in contrast to Einstein, who kept his political and scientific activities rigorously separate, Langevin tried to join the two. The developments in the immediate postwar period, especially the boycott of German scientists, which had widespread support in France, convinced him that he could serve both peace and science by working for the resumption of international scientific relations. In 1920, when he became editor of the *Journal de physique,* he strove to include articles and reviews of books by non-

---

[18] Ibid., p. 153; G. L. de Haas-Lorentz, *H. A. Lorentz. Impressions of his life and work* (Amsterdam, 1957), p. 148; Max von Laue to Wilhelm Wien, June 16, 1925, Wien Papers, DM.
[19] Heilbron, *Dilemmas of an upright man,* pp. 106–108.

Table 2. *Nominations made in favor of own-country candidates
by nominators of major countries, 1901–1933, in percentages
(N in parentheses)*

Nominator's country	Physics and chemistry	Physics	Chemistry
Germany	69 (591)	69 (307)	69 (284)
France	76 (306)	72 (181)	82 (125)
Great Britain	63 (91)	68 (60)	55 (31)
United States	65 (151)	64 (95)	67 (56)

French authors including German ones. His real tour de force in France, however, was his organizing Einstein's visit to Paris in 1922, the first one by a German scientist after the war. Internationally, he found his main field of action within the Solvay Physics Councils, where as a member of the scientific committee, he assisted Lorentz in his efforts to reopen the councils to German physicists.[20]

To what extent are the key dates and events indicated in this overview of international scientific relations during the 1914 to 1930 period reflected in nominations for the Nobel prizes in physics and chemistry? Put differently, were the nominations contingent on external events or were there pervasive factors that made for their isolation from social and political upheavals? The classification of each nomination by the nationality of the nominator and the nominee yields two separate indicators each of which sheds important light on this question.

The first indicator is the proportion of nominations made in favor of candidates from the nominators' own countries. In each of the four major powers – Germany, France, Great Britain, and the United States – own-country nominations represented between two-thirds and three-fourths of all nominations (see Table 2). Figure 1 gives an overview of the changes that occurred in the proportion of own-country nominations for each five-year period between 1901 and 1933. It is important to point out that, although no Nobel prizes in the sciences were awarded between 1915 and 1918, nominations continued to arrive in Stockholm, albeit in a diminished number. As indicated in Figure 1, during the war

---

[20] Bernadette Bensaude-Vincent, *Paul Langevin: Science et vigilance* (Paris, 1987), pp. 95–107.

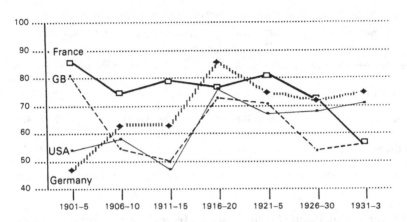

Figure 1: Own-country nominations by nominators from major countries for the Nobel prizes in physics and chemistry, 1901–1933 (in percentages).

the proportion of nominators in Germany, Great Britain, and the United States who restricted their nominations to their own compatriots increased sharply.

The timing of this upturn makes it reasonable to assume that the nominations were contingent on external events. The role played by the concurrent events of a surge of nationalism, particularly in Germany, and the breaking off of international scientific relations between countries at war will be examined in the next section. We will also address there the question of why the proportion of own-country nominations in the four major powers remained at a high level throughout the 1920s.

The second indicator is the nominations exchanged between scientists belonging to present or former hostile nations, that is, the votes given by nominators in the Allied nations to candidates belonging to one of the Central Powers, and vice versa. The next section will explore these patterns of exchange before, during, and after the war. That the two indicators are intimately linked becomes obvious when one considers that exchanges of votes only occur with respect to nominations that are not own-country ones. When own-country nominations rose (as happened during and after the war) to between 70 and 80 percent of the nominations made by scientists in most of the major powers, the proportion of nominations given scientists in other countries – be they allied, hostile, or neutral – necessarily diminished and sometimes became insignificant.

International exchanges of honors through Nobel nominations: a chronology of events

The golden age of internationalism (1900–1914)

The Nobel institution was founded on the internationalist ideals that reigned in science and culture at the turn of the century. Those ideals went beyond the belief that knowledge could transcend national boundaries proclaimed in the universalist creed. The speeches at the early Nobel ceremonies frequently expressed the hope that the prizes would foster an active internationalism in which scientists' work for the betterment of the human condition would turn these fields of endeavor into "battlefields for peace."[21] As conflicts multiplied in the relations among nations – in Manchuria, the Balkans, and North Africa – and among different social classes within nations, astounding progress was nevertheless being made, the speakers felt, in the fields of science, technology, and agriculture. Such progress had led to heretofore unimagined improvements in living conditions in which the general populations of all civilized nations had shared. This held out the hope, one speaker believed, that given enough time, the conditions making for combativeness would disappear and Alfred Nobel's dream for a peaceful evolution of mankind would be realized.[22]

The founders' hope that the nominators, although acting as representatives of their national scientific communities, would look further afield and put forth candidates whose support would thus have been international seemed initially to have been unrealistic. In the first five years, the number of nominations was small, between 25 and 40 per year for the physics prize and about the same for the chemistry one. The majority of these nominations were put forth in favor of candidates from the nominators' own countries; in the 1901 to 1906 period, this applied to 86 percent of the French nominations and 81 percent of the British ones (see Figure 1).

---

[21] *Les prix Nobel en 1911* (Stockholm, 1912), p. 9.
[22] *Les prix Nobel en 1903* (Stockholm, 1906), pp. 5–7; *Les prix Nobel en 1910* (Stockholm, 1911), pp. 8–10.

Critical and empirical studies

The favoring of one's own compatriots has already been discussed as endemic to the French nominators. In other countries, it may have resulted from the nominators seeing themselves as members of national juries who were to bring the foremost scientific achievement or scientist in their countries to the attention of the Swedish prize awarders. This interpretation is suggested by the spirited debate that broke out in the pages of the *Times* (London) and other journals after it was discovered that no Britisher was among the first year's award winners. This was attributed, among other things, to the complacency of the British nominators, who had not been sufficiently vigorous in defending the claims of British candidates. Various schemes were suggested to remedy the situation; the *Daily Telegraph* offered to advertise the names of the nominators so that worthy British candidates could be brought to their attention, and the British Society of Authors formed a Nobel Committee for the purpose of coordinating nominations for the literature prize.[23] Eventually, the number of British prizewinners increased and the debate abated.

For about ten years, between 1905 and 1915, the nominating system reflected the strong internationalism of the era. The proportion of own-country nominations dropped to an all-time low among British and American nominators and held steady among the German ones (see Figure 1). A larger number of votes thus went to scientists in other countries. When one charts the national origins of the largest proportion of nominations that each of the major countries received from the outside, one finds that Germany supplied the largest support for candidates in chemistry in both France and Great Britain; in physics, this was so for Great Britain.[24]

Scrutiny of some of the most active nominators in chemistry during this time reveals that they all belonged to a select group of internationally highly visible scientists. The members of this group of "stars" – Svante Arrhenius, Adolf von Baeyer, Emil Fischer, Henri Moissan, Wilhelm Ostwald, Wiliam Ramsay, T. W. Richards, and Richard Willstätter – were all Nobel prizewinners in the period 1901 to 1915.[25] As depicted

---

[23] *Times* (London), January 6, 1902; *Daily Telegraph*, January 8, 1902; *Westminster Gazette*, January 12, 1902.

[24] Crawford, *The beginnings of the Nobel institution*, pp. 106–108.

[25] Svante Arrhenius (1859–1927) was a bona fide member of this group. His nominations were skewed, though, as a result of his close involvement with the selection of Nobel prizewinners. Ibid., pp. 116–136.

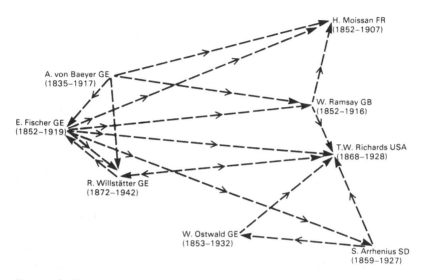

Figure 2: Nominations given and received among members of the international elite in chemistry, 1901–1915.

in Figure 2, the nominations they bestowed on each other make for an intricate web of relationships that cuts across national boundaries. Nominations were only one aspect of the multifaceted interactions of these men. Their biographies and especially the letters they exchanged show that they shared friendships, exchanges of honors other than Nobel nominations, and practical international scientific activities in both organic and inorganic chemistry. In most instances, their names were part of the show of support given the prizewinners of the early years – Fischer (1902), Ramsay (1904), von Baeyer (1905), Moissan (1906) – when they did not win awards themselves. German support for French chemistry candidates was considered normal enough for Paul Sabatier to attribute his own and Victor Grignard's Nobel prize in 1912 to such support. Being only guess work, this was wrong in Sabatier's case but correct for Grignard.[26] These nominating relationships reflected the intimacy of the international community of chemists in the prewar period. This small "world of chemistry," as it was often referred to by the Swedish prize

[26] Mary Jo Nye, *Science in the provinces: Scientific communities and provincial leadership in France, 1860–1930* (Berkeley, 1986), pp. 141–142, 150, 184–185.

jury fretting over how its award decisions were received, would change dramatically as a result of World War I.

The war and its aftermath (1916–1920)

In 1915 and 1916, 100 percent of the German nominations were made in favor of German candidates, whereas in the two previous years, the corresponding figures were 64 and 70 percent. For the prizes of 1916, Marcel Brillouin, a French mathematical physicist, outdid his compatriots in own-country nominations by single-handedly proposing 13 French physicists, some of whom were at the front.[27] For the British nominators, by contrast, wartime preoccupations meant that hardly any nominations were made, a pattern that was to continue throughout the war. During the war, the nominations made by scientists in the belligerent nations rarely went to candidates on the opposite side. As indicated in Figures 3 and 5, for both Central Power (primarily German and Austrian) and Allied candidates, the portion of the vote received from enemy countries dropped from more than 10 percent in the period 1911 to 1915 to 2 percent during the war and its immediate aftermath.

The vigor with which German nominators joined forces around their own candidates was at one with the patriotic surge – the famous spirit of 1914 – that swept through the universities early in the war. Neither in France nor in England did scientists join the war effort on the propaganda front with the total commitment and enthusiasm with which German academics, almost without exception, answered the call to arms. The two manifestoes referred to in the previous section are the most striking examples of how scholars put their good names and their reputation for objectivity in the service of the nation at war. In both, they refuted the claims that German forces had committed atrocities in Belgium, and declared themselves to be at one with the army, "whose *Geist* is that of the German *Volk*."[28] Among the signers of the Appeal of the

---

[27] For the names of Brillouin's nominees, see Elisabeth Crawford, J. L. Heilbron, and Rebecca Ullrich, *The Nobel population, 1901–1937: A census of the nominators and nominees for the prizes in physics and chemistry* (Berkeley and Uppsala, 1987), p. 65.

[28] Böhme, *Aufrufe und Reden*, p. 49.

Ninety-three Intellectuals were all the German Nobel laureates in chemistry whose international network we discussed earlier.

More in private, the signers of the manifestoes bombarded their colleagues in neutral countries – "enemy" scientists being of course inaccessible – with propaganda statements that they often wanted publicized. This tried the patience of even the most well-meaning and neutral scientists. When Lorentz tried to "educate" Wien about the theses *"Es ist nicht wahr,"* that is, the denial of German war crimes, in the Appeal of the Ninety-three Intellectuals by sending counter-propaganda from France and Belgium, Wien turned the literature over to the Prussian Ministry of War. Lorentz promptly broke off correspondence. Likewise, when Arnold Sommerfeld tried to enlist the Dutch physicist and Nobel prizewinner Heike Kamerlingh Onnes in the campaign to transform the University of Ghent in German-occupied Belgium from a predominantly French-speaking to a Flemish institution, he was sharply told not to interfere in the linguistic policies of another country.[29]

Naturally, there were parallel activities carried out by scientists in Allied countries. Outrage over German atrocities filled their declarations in public and private. Whereas some of these atrocities were imagined, most of them were real enough with none causing as much outcry as the use of poison gas during the battle of Ypres in 1915.[30] When feelings were running particularly high even the honors that scientists of the now belligerent nations had bestowed on each other in the past were revoked or in danger of becoming so. Early in 1915, the Academy of Sciences of Paris decided to strike out the names of the signers of the Appeal of the Ninety-three Intellectuals it had elected to foreign membership. In

---

[29] H. A. Lorentz to Wilhelm Wien, March 22, 1915, and May 3, 1915, Wien Papers, DM; Heike Kamerlingh Onnes to Arnold Sommerfeld, June 6, 1916, Sommerfeld Papers, DM; Michael Eckert et al., *Geheimrat Sommerfeld – Theoretischer Physiker: Eine Dokumentation aus seinem Nachlasss* (Munich, 1984), p. 130; Fritz Haber to Svante Arrhenius, September 23, 1914, Arrhenius Collection, KVA.

[30] Badash, "British and American views," pp. 102–108; L. F. Haber, *The poisonous cloud: Chemical warfare in the First World War* (Oxford, 1986). Ernest Rutherford's statement (in a letter to Arrhenius) in 1915: "Germany would employ any method of warfare however barbarous" is only one of many similar views expressed in private. Rutherford cites the use of gas warfare and the poisoning of drinking water in France and concludes: "I am expecting the use of even worse things in the future." (Ernest Rutherford to Svante Arrhenius, June 1, 1915, Arrhenius Collection, KVA.)

the Academy of Sciences of Berlin and the Royal Society of London, moves to exclude "enemy scientists" or to break off relations once peace had come were only narrowly defeated.[31] These and other events gave scientists on both sides of the conflict the feeling that relations had been irreparably damaged. There were some differences, however, for in contrast to their colleagues in Allied countries, some prominent German scientists felt that their own country provided such fertile soil for science that they could do without knowledge of the work carried out in enemy countries.[32]

In view of these feelings of self-sufficiency and the arrogance they bred, it is surprising that, on the whole, German scientists went out of their way to keep the Nobel institution out of the strife, even though, as has already been shown, many Nobel prizewinners were active on the propaganda front. When it was announced in the fall of 1915 that Max von Laue and W. H. and W. L. Bragg had won the prizes in physics for 1914 and 1915, respectively, for their work in X-ray crystallography, von Laue "sincerely regretted that the ongoing world war prevented him from personally wishing the Braggs joy." After his well-wishes had been transmitted to the Braggs through the good offices of Svante Arrhenius and had been reciprocated, von Laue wrote and expressed his delight that "such unusually successful researchers wanted to congratulate me despite the war."[33] The same concern for keeping the prizes out of the strife is hinted at by their not being used publicly as pawns in the propaganda war. In contrast with the Royal Society's prestigious Rumford medal, which Lenard and Röntgen (who had won it jointly in 1896) donated to the Red Cross with considerable fanfare, von Laue's use of the Nobel prize to buy victory bonds was not publicized. This despite the fact that von Laue was no less convinced than Lenard and Röntgen of his "duty to follow his *own* conviction," as he wrote in a

---

[31] Badash, "British and American views," pp. 109–112; Heilbron, *Dilemmas of an upright man*, pp. 79–80.

[32] Philipp Lenard, *England und Deutschland zur Zeit des grossen Krieges* (Heidelberg, 1914), pp. 15–16; Wilhelm Wien to C. W. Oseen, November 19, 1914, cited in Wien, *Aus dem Leben und Wirken*, p. 60; Wilhelm Wien to Philipp Lenard, June 19, 1916, Wien Papers, DM.

[33] Max von Laue to Svante Arrhenius, November 30, 1915, March 10, 1916; Wilhelm Wien to Svante Arrhenius, January 8, 1916, Arrhenius Collection, KVA; see also Max von Laue to Wilhelm Wien, November 14, 1915; Wilhelm Wien to Max von Laue, November 15, 1915, Wien Papers, DM.

letter to Arrhenius, and "not feel any moral scruples about using the prize money to buy war bonds."[34]

The caution that German scientists had taken to insulate the Nobel institution from their propaganda war probably helped in making it possible for all the five Germans, among them Max von Laue, Max Planck, and Fritz Haber, whose prizes had accumulated during the war to attend the first postwar prize ceremony held in 1920. The ceremony was probably one of the first occasions after the war for an official meeting between German and Allied scientists. At this time, the exclusion of German scholars from international meetings was only the outward manifestation of a rupture between former enemy physicists and chemists that was almost total. In any event, international meetings could hardly be organized at a time when runaway inflation, transport problems, food shortages, and visa restrictions afflicted the populations of all the former belligerent nations in Europe. No wonder even the neutral scientists who were ready to work for resuming some relations through informal contacts despaired and concluded that for the time being they could do no more than wait and see.[35]

The prize ceremony also contained many of the elements that would make the reconstruction of international science so painfully slow. There was the presence of Haber, who had led the German chemical warfare operations, and whose 1918 award (citing his prewar work on the synthesis of ammonia from its elements) seemed particularly ill-advised in the eyes of many Allied scientists. There was the absence of T. W. Richards, the American chemist and undisputed international expert on atomic weights, who had begged off, giving as his excuse what eventually became the refrain of Allied scientists that contact could not be reestablished until the signers of the Appeal of the Ninety-three Intellectuals recanted. Or, as Richards put it to Arrhenius, "Apology to the civilized world is necessary, for their extraordinary manifesto was addressed to the civilized world."[36] That the prize ceremony nevertheless was held and that the awards resumed their normal course in the 1920s were in

---

[34] Badash, "British and American views," p. 106; Lenard, *England und Deutschland*, 4n; Max von Laue to Svante Arrhenius, March 19, 1916, Arrhenius Collection, KVA.

[35] Philippe Guye to Svante Arrhenius, February 23, 1919, February 21, 1920; Ernst Cohen to Svante Arrhenius, December 23, 1919, Arrhenius Collection, KVA.

[36] T. W. Richards to Svante Arrhenius, February 12, October 8, December 16, 1919; February 16, 1920, Arrhenius Collection, KVA.

large measure due to the Nobel prizes being the unassailable symbol of an internationalism in science that many felt was a casualty of the war.[37]

Reconstruction through informal contacts (1921–1925)

The vignettes of Planck, Wien, and Langevin show that in the first half of the 1920s, Allied and Central Power scientists gradually and hesitantly tried to find ways to resume prewar relations. The public and official route to reconciliation was barred by the continued refusal of the IRC to admit the Central Powers. Although the disciplinary associations, members of the IRC (of which there were half a dozen by 1925), more and more used their own judgment as to whether or not former enemy scientists would be excluded from their meetings, by mid-decade, such exclusions still applied to about 50 percent of the meetings.[38]

To the scientists on both sides who were concerned about reconstruction, informal and unofficial contacts appeared, then, not just to be a reasonable way of rebuilding international science but the only way. These contacts ran the gamut from a simple exchange of greetings, using the good offices of a neutral, as in the case of Georges Urbain, the French inorganic chemist and his former collaborator and friend R. J. Meyer in Berlin, to the elaborate plans that were laid for a "little unofficial international chemists' meeting" in neutral Netherlands by F. G. Donnan, the successor to Ramsay's chair at University College, London, and Ernst Cohen of the University of Utrecht. The meeting of about 40 chemists from all the major powers except France, and from many other countries as well, took place in Utrecht in the summer of 1922 the week before the International Union of Pure and Applied Chemistry held its meeting in Lyon from which scientists representing the former Central Powers were excluded.[39] One is struck by the fact

---

[37] In his welcoming address at the 1920 prize ceremony, the president of the Nobel Foundation, Henrik Schück, struck an internationalist note when he saw the celebration as an expression of a hope "that will never be allowed to die: That in the end science and literature will burst the cloud of hatred between people that has invaded the world." (*Les prix Nobel en 1919–1920* [Stockholm, 1922], p. 13.)

[38] Grundmann, "Zum Boykott," p. 800.

[39] Georges Urbain to Svante Arrhenius, (?) 24, 1922, November 18, 1922; Ernst Cohen to Svante Arrhenius, May 25, 1921; F. G. Donnan to Svante Arrhenius, February 1, 1921, November 24, 1921, February 20, 1923, Arrhenius Collection, KVA; Svante Arrhenius to F. G. Donnan, February 9, December 3, 1921; December 2, 1923, F. G. Donnan papers, University College, London.

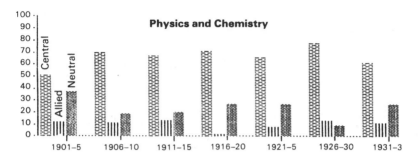

Figure 3: Proportion of nominations received by Central Power candidates from nominators in Central Power, Allied, and neutral countries, 1901–1933 (in percentages). "Allied Nations" = Belgium, Canada, England, France, Italy, Russia, and the United States. "Central Powers" = Austria, Czechoslovakia, Germany, Yugoslavia, Hungary, and Poland. "Neutral" countries = Denmark, Sweden, Switzerland, Norway, Finland, Spain, and the Netherlands.

that those initiating the contacts were motivated less by a strong need for practical international cooperation in the area of communication of research results, for instance, than by a sincere desire to show the way in an effort to reconstruct not just international science but other international cooperative efforts as well.[40]

In view of these developments, it is surprising to find that there was relatively little change in the wartime pattern of nominations for the Nobel prizes in physics and chemistry once peace had returned. As indicated in Figure 1, the proportion of own-country nominations remained high for all the major powers in the period 1921 to 1925: between 70 and 80 percent. The wartime pattern also held more or less for the nominations that candidates from the former Central Powers received from Allied scientists (see Figure 3). Those nominations remained at 8 percent or well below the prewar level of 13 percent.

On the surface, then, the Nobel nominations appeared to be more in tune with the boycott and counterboycott than with the conciliatory

---

[40] F. G. Donnan wrote to Svante Arrhenius about the Utrecht meeting in the following terms: "Let our meeting be an occasion – indeed a solemn occasion – for the rebirth of peace and goodwill amongst men. If science does not make a great effort to lead in this work, then there will be little hope for the world." (F. G. Donnan to Svante Arrhenius, November 24, 1921, Arrhenius Collection, KVA.)

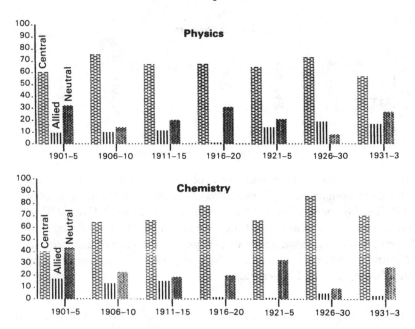

Figure 4: Proportion of nominations (tabulated separately for physics and chemistry) received by Central Power candidates from nominators in Central Power, Allied, and neutral countries, 1901–1933 (in percentages). For explanations of the terms Allied, Central Power, and neutral nations, see Figure 3.

moves initiated by neutral scientists. When the nominations are tabulated separately for physics and chemistry (as shown in Figure 4), important differences between the two disciplines emerge. In 1921 to 1925, Allied nominations of Central Power physicists rose to 14 percent, whereas those of chemists remained on the wartime level of 1 percent. This increase in physics was due almost entirely to the candidacy of Albert Einstein, who doubtless was Germany's greatest asset in its efforts to regain a position in international science in the early 1920s.[41] Once Planck had won his Nobel prize, Einstein became the most nominated candidate until 1922 when he received the unawarded prize for 1921.[42]

[41] Heilbron, *Dilemmas of an upright man*, pp. 115–116.
[42] Elisabeth Crawford and Robert Friedman, "The prizes in physics and chemistry in the context

Einstein's candidacy lies behind the sharp divergence between physics and chemistry that ensued in the period 1921 to 1925.

Einstein's scientific stature, the admiration many felt for his special theory of relativity, and his outspoken pacifism during the war were clearly the most important factors in the support he received from British, French, and American nominators. That the politics of reconstruction also played a role is suggested not only by the strong presence of scientists from neutral countries among the nominators – in 1920, the three Dutch laureates in physics led by H. A. Lorentz – but, more important, by that of the French physicists (Langevin, Marcel Brillouin, and Jacques Hadamard), who had received Einstein when he came to lecture at the Collège de France in 1922. This visit, which was the first public appearance by a German scientist since the war, was organized by Langevin in the same spirit that had given rise to other informal encounters. Following Einstein's wishes, the visit was kept almost exclusively scientific. The esoteric nature of the discussions also provided the public, which was kept "informed" by the popular press, with an image of high intellect capable of transcending the boundaries of language and nationality while at the same time, as personified by Einstein, remaining intensely humane.[43]

Relative normalization (1926–1933)

By the mid–1920s, international scientific relations had been largely restored for each of the three levels outlined earlier: governmental, international organizational, and personal. With respect to the first two levels, a change in the statutes of the International Research Council, which lifted the bars against the admission of the Central Powers, was instituted in 1926. Because Germany refused to join, this action was mainly symbolic. In any event, German scientists now participated in international meetings, whether or not German academies and scientific societies were members of the IRC. In 1926, German scholars were excluded from only about 15 percent of international meetings with the exclusions continuing to drop off until 1932 when they seemed to have

of Swedish science," in Carl Gustaf Bernhard, Elisabeth Crawford, and Per Sörbom, eds., *Science, technology and society in the time of Alfred Nobel* (Oxford, 1982), pp. 323–324.
[43] Michel Biezunski, "Einstein à Paris," *La recherche* 13 (1982): 502–510; Bensaude-Vincent, *Paul Langevin*, pp. 99–104.

reached zero.[44] Personal exchanges between scientists from the former Allied and Central Powers also returned to normalcy as more and more scientists traveled the routes indicated in the thumbnail sketches of Planck, Wien, and Langevin.

These developments did not significantly affect the patterns of nominations for the Nobel prizes, which continued to exhibit many of the same features of the 1921 to 1925 periods. The proportion of own-country nominations remained high; only those of French nominators declined sharply in the early 1930s (see Figure 1). The increase in the nominations that crossed former enemy lines was most apparent for Allied candidates, who now received 18 percent of their votes from nominators representing the Central Powers as compared with 8 percent in the period 1921 to 1925 (see Figure 5). However, this was still below the prewar figure of 25 percent. More important, the differences between physics and chemistry continued. In the period 1921 to 1925, Central Power physicists received 14 percent of their nominations from Allied scientists; this figure rose to 20 percent in the years 1926 to 1933. In the same periods Central Power chemists received 1 percent of their nominations from Allied scientists, which increased to 4 percent (see Figure 4). There was less of a difference between the two disciplines in the nominations going from Central Power to Allied scientists. In the period 1926 to 1933, 18 percent of the vote for Allied physicists came from former enemy scientists as compared with 12 percent of that for Allied chemists (Figure 5).

When one looks behind these figures, one finds that specific circumstances rather than a general inclination continued to dictate Allied support for the candidacies of German physicists. Once Einstein had received the prize (in 1922), Allied nominators turned their attention to the next generation of German theoretical physicists. The major part of the Allied vote went to Werner Heisenberg and Erwin Schrödinger, who were awarded the Nobel prizes in 1932 and 1933, respectively, for their contributions to quantum mechanics. A less important part concerned atomic scientists (Peter Debye, Walter Gerlach, and Otto Stern). These votes point to the very special conditions required to fill the void that the war had caused in international exchanges of honors: the emer-

---

[44] Schroeder-Gudehus, *Les scientifiques et la paix*, pp. 262–289; Kevles, "'Into hostile political camps'" p. 60; Grundmann, "Zum Boykott," p. 800; "Deutsche Wissenschaft und Ausland," p. 331.

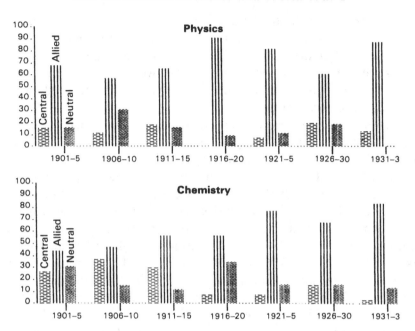

Figure 5: Proportion of nominations received by Allied candidates from nominators in Central Power, Allied, and neutral countries, 1901–1933 (in percentages). For explanations of the terms Allied, Central Power, and neutral nations, see Figure 3.

gence of an international specialty community, which in this case was built around the fields of atomic physics and quantum mechanics; the preeminent contribution of German physicists to the development of knowledge in these fields; and, perhaps most important, the politically noncontroversial nature of the research.

Exchanges of research results in these fields were no doubt facilitated by the rapid revival of the Solvay councils. The first postwar council was held in 1921. The reopening of the council to representatives of the Central Powers was undertaken, timidly, in 1925 when only Schrö-dinger attended, and decisively in 1927, when Born, Einstein, and Planck participated.[45] For this to occur had required the patient efforts of two

---

[45] In contrast with the Solvay councils in physics, the first and second Solvay councils in chemistry

successive presidents, both with a strong internationalist outlook: H. A. Lorentz, who was instrumental in reestablishing the councils on a broader basis after the war, and Paul Langevin, who took over upon Lorentz's death in 1928 and organized the two councils in the early 1930s that confirmed the importance of these gatherings for exchanges in atomic and theoretical physics.[46]

In chemistry, the void caused by the war went much deeper than in physics. Between 1915 and 1927, no Allied scientist nominated a German chemist, and when this finally occurred, the nomination concerned Peter Debye, an atomic scientist with Dutch nationality (professor of experimental physics at the University of Leipzig), who had already been proposed several times for the physics prize. Chief among the reasons for the nominators' resistance to German chemists was undoubtedly a lingering feeling of revulsion against a profession whose leading members – Haber and Walther Nernst, but also Otto Hahn and Heinrich Wieland, all Nobel prizewinners after the war – had initiated the use of poison gas.[47] That Allied nominators shunned the chemists associated with the Kaiser-Wilhelm Society, where Haber's Institute for Physical Chemistry had been the army's center for chemical warfare, supports such an interpretation.

German and Allied scientists may also have been less likely to exchange votes because of the climate of national competition and secretiveness created by the inroads that industrial research and development was making into the basic research sphere. The Nobel prizes, which heretofore had almost exclusively been awarded academic scientists, were also affected by those developments. In the early 1930s, not one but three industrial chemists received prizes: in 1931, Carl Bosch and Friedrich Bergius of I. G. Farben and, in 1932, Irving Langmuir of General Electric. Still other explanations relate to the demise of the international elite of chemistry "stars" who exchanged honors across both national and specialty boundaries during the golden age of internationalism. What little remained of this network after the estrangements

---

held, respectively, in 1922 and 1925 had no German participation. This did not occur until the third council (1931), which was attended by W. Schlenk (Berlin) and H. Staudinger (Zurich).

[46] Bensaude-Vincent, *Paul Langevin*, pp. 142–147; Heilbron, *Dilemmas of an upright man*, pp. 107–108.

[47] Haber, *The poisonous cloud*, passim. So many more prominent German academic scientists than Allied ones took part in gas warfare that had the Nobel committees required noninvolvement as a condition for receiving the prize, they would have had to eliminate many valid candidates.

caused by the war disappeared in the postwar period with the deaths of Arrhenius and von Baeyer in 1927, Fischer in 1919, Ramsay already in 1916, and T. W. Richards in 1928, to mention only the most prominent members of the network.

As indicated in Figure 5, Central Power, predominantly German, nominators were only marginally more inclined to give their support to Allied scientists in the period between 1926 and 1933 than vice versa. However, the fact that they did so at all, and especially without the stimulant of a candidate such as Einstein, suggests that they were more in tune with the conditions of relative normalization that ensued after 1926 than were their Allied counterparts. The nominations data and other materials provide scattered evidence of this and how it came to be. First, whereas German physicists concentrated their nominations in atomic physics (A. H. Compton, R. A. Millikan, and R. W. Wood), German chemists were not only more likely to put forth Allied candidates than vice versa but were also more eclectic in their choices. Second, German nominators were, on the whole, not falling in with the counter boycott and its ban on the resumption of relations with the victors, urged by such hard-line organizations as the Verband der Deutschen Hochschulen (VDH). That the Nobel institution continued to be insulated from the politics of science probably helped; for example, in the period 1920 to 1933, the *Mitteilungen* of the VDH did not once use the award of Nobel prizes to Germans for propaganda purposes.[48]

A third and final explanation for the larger propensity of German nominators to give their votes to Allied scientists may have been the importance that German academics in the Weimar period attached to their country's regaining its hegemony in science and culture. In the opinion of Paul Forman and Brigitte Schroeder-Gudehus scientific achievements became the surrogate (*Macht-Ersatz*) for the other lost attributes of a great power.[49] Because recognition from abroad was seen as essential to great power status in the sciences, it fostered a kind of scientific internationalism. The fact that German physicists and chemists won as many Nobel prizes in the decade and a half following the war as they did in the one preceding it, and that these awards reflected important breakthroughs in atomic physics and quantum mechanics,

---

[48] I am grateful to Bernhard vom Brocke for having searched this journal for me.
[49] Forman, "Scientific internationalism and the Weimar physicists," pp. 161–165; Schroeder-Gudehus, *Les scientifiques et la paix*, pp. 223–239.

confirmed the view that Germany was still a great scientific nation.[50] In this respect, the Nobel prizes provided important support for the resumption of international scientific relations.

Conclusion and discussion

What do the patterns in the nationalist and internationalist orientations of the Nobel nominators revealed here contribute to our understanding of nationalism and internationalism in science? It can be concluded, first of all, that the disturbances in international scientific relations caused by World War I varied, depending on the level of interaction; hence, the IRC is not sufficient as a case to test the consequences of the war. It remains puzzling that while most activities in the different spheres of interaction returned to relative normalcy after 1926, exchanges of honors through Nobel nominations continued to reflect the patterns set by the war well into the 1930s and probably even later.[51]

But was not the chauvinism of the Nobel nominators during and after the war a peculiarity of the university-based elite rather than common to scientists at large? A positive answer to this question would be in line with Cock's argument that the nationalist and vindictive policies instituted against the former Central Powers after the war were the work of the scientific elites that dominated the IRC and its member organizations. In Cock's opinion, had the generality of scientists been consulted on these matters, the "natural" tendencies of scientists toward universalism would have come to the fore, and international scientific relations would have been resumed more rapidly.[52]

The question cannot be answered directly. An indirect answer emerges, however, when we divide the Nobel population into two groups and compare them with respect to the degree of nationalist bias shown in their nominations. The former group comprised the large majority

---

[50] See, among many other comments, those of Haber, who on learning that he had received the Nobel prize wrote to Arrhenius that "the entire German nation feels concerned by this honor" (Fritz Haber to Svante Arrhenius, December 5, 1919, Arrhenius Collection, KVA); Pieter Zeeman, a pro-German Dutchman, found that the awards of 1919 confirmed "the victory of German science that eventually everybody will admit" (Pieter Zeeman to Arnold Sommerfeld, January 16, 1920, Sommerfeld Papers, DM).

[51] The same picture would probably emerge if one studied other marks of honor such as election to foreign membership in academies of science, honorary doctorates from leading universities, and invitations to foreigners to deliver prestigious lectures at these universities.

[52] Cock, "Chauvinism and internationalism," pp. 275–277.

of nominators, who by virtue of their positions, constituted a university-based elite. The latter was made up of nominators who had already won the prizes in physics and chemistry. This distinction had given them a permanent franchise that a large number of them exercised. (For the overall period of 1901 to 1933, about one-third of the nominations in physics and chemistry were handed in by those who had already won the prize.)

On the whole, the nominations made by the prizewinners did not show more or less chauvinism than those of the nominators who were not prizewinners. One way to gauge chauvinism is to look at the proportion of votes that Allied nominators bestowed on Central Power candidates and vice versa in the sensitive period 1916 to 1933. Thirteen percent of the vote of nonprizewinners from both camps was exchanged, 10 percent of that of the prizewinners. This difference can hardly be considered large enough to turn the latter into a group of chauvinists. More important, there was no difference between the elite and the ultra-elite with respect to the influence of the watershed of 1915 over their voting patterns. From the first year of the prizes until 1915, one-fifth of the vote of the nonprizewinners and one-fourth of that of the winners was exchanged in the manner indicated; from 1916 to 1933, such exchanges concerned only about 10 percent of the overall vote.[53]

The more nationalist orientations of prominent members of the ultra-elite after 1915 can be illustrated by Planck, who, starting in 1901, put forth candidates in physics, chemistry, or both for 23 of the 33 years covered here. This record was broken only by Emil Warburg, director of the Imperial Institute of Physics and Technology (Physikalisch-Technische Reichsanstalt) at Berlin-Charlottenburg, who nominated in 30 of the 33 years. In the prewar era, neither Planck nor Warburg gave any hint of nationalism in their nominations, many of which concerned Anglo-Saxon scientists: Lord Rayleigh, Ernest Rutherford, J. J. Thomson, and in 1914, on the eve of the war, W. H. Bragg. After the war, Planck, who in so many other ways sought to renormalize international scientific relations, only once put forth a scientist from a former Allied nation (A. H. Compton in 1927); Warburg never did so. Indeed, like

---

[53] When comparing the periods before and after World War I, one has of course to take into account that the Nobel prizewinners were more numerous in the latter period and therefore made more nominations (316 out of a total of 887 in the period 1916–1933 as compared with 186 out of 608 in the years 1901–1915).

many of his colleagues at the Kaiser-Wilhelm Society, Planck tended to favor candidates from that institution. (See Chapter 5.)

Surprising as it may seem, other German members of the ultra-elite with a much stronger nationalist bent were more adventuresome: Stark, the great promoter of *"Deutsche Physik,"* for instance, who nominated Rutherford for a second prize, this time in physics in each of the years 1931, 1932, 1933, 1935 and 1937; or Wien, who made the first postwar German nomination of an Allied scientist in 1925 when he included R. W. Wood in a proposed three-way split of the prize that also involved Lenard and Friedrich Paschen. (In the following year, Lenard returned the honor by nominating Wien alone for a second prize in physics.)

In conclusion, then, when it came to exchanges of honors, internationalism in science was indeed a casualty of the war. But apart from the disillusionment that this may cause those who see internationalism as inherent to science, did it really matter? Nationalism was a pervasive feature of the nominating system, and this partly by design, as the nominators were expected to act as representatives of their national scientific communities. The role of nationalism as a force stimulating competition among national scientific enterprises carried the danger, however, that it might produce pathologies in the form of unfair competition or secrecy. The chauvinism of nominators from major powers during and immediately after the war was in many ways pathological, although it had only limited consequences for the Nobel institution. Nevertheless, it seems doubtful that the Nobel Committee for Chemistry would have taken the decision to award the prize to Haber so soon after the war had there been a semblance of international support for other candidates.[54] One also wonders to what extent the Nobel Committee for Physics would have deviated from the course it took through its string of awards in atomic and theoretical physics in the interwar period had there been other fields with the same number of nominations exchanged across national borders. These conjectures do not modify the viewpoint that the symbolic value of the prizes was not diminished by the upheavals in international science caused by World War I.

---

[54] In the opinion of one observer, the chemistry committee discussions of Haber's award in 1918 and 1919 represented "a curious mixture of ignorance and irrelevance." Haber, *The poisonous cloud*, p. 392n.

4

~~~~~~

Center-periphery relations in science: the case of Central Europe

The focus here is on Joseph Ben-David's attempt to use the center-periphery dichotomy to explain the dynamics of scientific development. As we have seen in Chapter 1, according to Ben-David's model, the countries that became scientific centers in modern times were those where the organizational structure for research was built on competition. This produced the innovations that raised the level of scientific activity not just in the country that had taken the lead but generally. Smaller countries constituted the periphery because for various reasons, mainly linguistic ones, they could not compete internationally with the organizational units at the center. All they could hope to do would be to copy the organization of scientific work at the center and thereby adopt its work orientations. In both these respects, however, the center would always retain a monopolistic position.[1]

During the first third of the twentieth century, east-central Europe (defined here as the Austro-Hungarian Empire and its successor, the nation-states of Austria, Hungary, and Czechoslovakia) was made up of the small and, in the case of Hungary and Czechoslovakia, linguistically marginal, scientific communities that would make the region peripheral according to Ben-David's use of the term. But these countries were also part of Central Europe, which put them in close proximity both geographically and linguistically to Germany, at the time the scientific center of the world. East-central Europe possessed a well-developed, time-honored system of higher education, which in many respects was modeled on that of Germany. That the basic organizational structures for

[1] See Chapter 1 and Joseph Ben-David, *The scientist's role in society: A comparative study* (Englewood Cliffs, N.J., 1971), pp. 171–173.

79

research were so similar to the German ones makes it easier to identify work orientations and organizational arrangements that were specific to the southern perimeter of Central Europe. In this way, one will, hopefully, arrive at a more nuanced view of center-periphery relations than that put forward by Ben-David.

In the following discussion, four different aspects of science in east-central Europe that represent a different type of relationship with the center are examined: institutional arrangements for teaching and research, the intellectual innovations of peripheral scientists, the recognition given their work at the center, and the emergence of specialties and work orientations specific to the periphery. These are investigated using the population of 81 Nobel nominators and nominees for the prizes in physics and chemistry in Austria, Hungary, and Czechoslovakia in the period 1901 to 1939. The majority (66 individuals) were chairholders at the universities in the region who had answered the invitation of the Royal Swedish Academy of Sciences to nominate candidates for the prizes in physics and chemistry in a given year.[2] There were 17 candidates, 4 of whom became prizewinners. (The candidates' names, as well as the universities and *technische Hochschulen* from which the Nobel population was drawn, are listed in Tables 3 and 4.)

All in all, the population comprised about two-thirds of the ordinary and extraordinary professors at universities and *technische Hochschulen* from about 1880 to 1939.[3] Because they were the ones who gave direction to both teaching and research, they are well suited to characterize physics and chemistry in the area during the period. As will be apparent, members of the population were also highly productive researchers.

The following types of data about the population have been collected to address the four aspects of relationship with the center:

1. The dependence on the center with respect to institutional arrangements for teaching and research has been studied using

---

[2] The Austrian universities and *technische Hochschulen* (Vienna, Graz, and Innsbruck) were invited with greater frequency than those of the other east-central European countries. Mainly for this reason, there are fifty-five Austrians among the nominators and nominees as compared with twenty Czechs and six Hungarians.

[3] *Minerva. Jahrbuch der Gelehrten Welt*, vols. 1–28 (Strasbourg, Berlin, and Leipzig, 1891–1932). The universities often interpreted "chairholders" broadly to include extraordinary and emeritus professors as well as those having only the title of professor with the result that the nominators sometimes represented several academic generations.

Table 3. *Candidates for the Nobel prizes in physics and chemistry from east-central Europe, 1901–1939*

	Nobel prize
Carl Auer von Welsbach (1858–1929)	
Ludwig Boltzmann (1855–1906)	
Joseph Maria Eder (1855–1944)	
Felix Ehrenhaft (1879–1952)	
Friedrich Emich (1860–1940)	
Loránd von Eötvös (1848–1919)	
Julius von Hann (1839–1921)	
Viktor Hess (1883–1964)	Ph 1936
Jaroslav Heyrovsky (1890–1967)	Ch 1959
Viktor Kaplan (1876–1934)	
Ernst Mach (1838–1916)	
Fritz Pregl (1869–1930)	Ch 1923
Erwin Schrödinger (1887–1961)	Ph 1933
Hans Zdenko Skraup (1850–1910)	
Ernst Späth (1886–1946)	
Siegmund Strauss (1875–?)	
Emil Votocek (1872–1950)	

Source: Elisabeth Crawford, J. L. Heilbron, and Rebecca Ullrich, *The Nobel population, 1901–1937: A census of the nominators and nominees for the prizes in physics and chemistry* (Berkeley and Uppsala, 1987).

data on the training and career trajectories of the entire population.

2. The innovativeness of work is revealed in the research orientations of the 17 individuals who were candidates for the Nobel prizes.

3. The recognition given the work of peripheral scientists at the center is determined using publication and citation measures derived from the *Physics citation index, 1920–1929.*

4. The emergence of work orientations and research styles specific to the periphery is examined drawing on biographical data concerning the members of the circle around Franz S. Exner, professor at the University of Vienna, whose students (12 of whom figure as members of the Nobel population) dominated physics in the region during the first third of the century.

Critical and empirical studies

Table 4. *Universities and* technische Hochschulen *represented among the Nobel population in east-central Europe*

Universities	Technische Hochschulen
Budapest (1784)	Brünn German (1849)
Czernowitz (1875)	Brünn Czech (1899)
Graz (1585)	Budapest (1856)
Innsbruck (1474)	Graz (1811)
Prag German (1348)	Prag German (1806)
Prag Czech (1882)	Prag Czech (1868)
Vienna (1365)	Vienna (1815)

Note: Date in parentheses is the year founded.
Source: Minerva. Jahrbuch der Gelehrten Welt, Vol. 1–28 (Strasbourg, Berlin, and Leipzig, 1891–1932).

Institutional arrangements for teaching and research in physics and chemistry

Teaching and research in physics and chemistry took on their modern forms in the universities of the Austro-Hungarian Empire in the latter part of the nineteenth century. It resulted in structures that gave preeminence to the chairholder, who set curricula and directed research. Research was usually carried out in an institute, comprising laboratories for students and faculty, attached to the chair. This organization of teaching and research, as well as the ideology of *Lernfreiheit* and *Lehrfreiheit* that accompanied it, was that of the German universities. As in Germany, the recognition of new disciplines in universities with long-standing traditions of learning – the university in Prague was founded by Emperor Charles IV in 1348 and the one in Vienna in 1365 – came about through the creation of chairs. The same mechanism was used to accommodate the new specialties spun off from larger disciplinary entities: physical chemistry, theoretical physics, and applications of physics and chemistry in technology.

With the growth in the number of students and postulants for chairs, physics and chemistry posts came to form the same hierarchical system as that found at German universities. On the bottom were the assistants and *Privatdozenten*. The former ran laboratory courses for students and the latter gave instruction for a fee. Further up the career ladder were the extraordinary professors, some of whom would receive only the title and no salary. At the top were the ordinary professors, who most often were also directors of the research institutes or laboratories attached to the chair. Early in the twentieth century, this system was well in place in the universities and *technische Hochschulen* of the Austro-Hungarian Empire.

Some 35 physicists, almost all ordinary or extraordinary professors, were among the 80 members of the Nobel population in east-central Europe.[4] Four-fifths of these were at the universities. As in the German universities, the main chair of physics was most often held by an experimentalist, who was also the director of the physics institute. Among the candidates for the Nobel prize who held these chairs were Ernst Mach at the German-speaking Prague University (1867–1895), Loránd von Eötvös at Budapest (1896–1909), Felix Ehrenhaft at Vienna (1920–1938), and Viktor Hess at Innsbruck and Graz (1925–1938).

In the late nineteenth century, theoretical and mathematical physics emerged as independent specialties and were given recognition through the creation of new positions. At first, these were often extraordinary or titular professorships. Subsequently, their incumbents were elevated to the rank of full professor and given their own institute, referred to as the "second physics institute." The peregrinations of Ludwig Boltzmann between the universities of Vienna and Graz (with side trips to those of Munich and Leipzig) between 1869 and 1902 illustrate the slow process through which theoretical physics gained recognition. Not until 1902 did Boltzmann finally find himself both with a chair and his own institute, in Vienna.[5]

There were about as many chemists as physicists in the population,

---

[4] The counts of academic physicists in Forman, Heilbron and Weart, "Physics circa 1900," show a total of thirty-five ordinary and extraordinary professors at the universities of the Austro-Hungarian Empire in 1900 and about the same in 1910. The higher number compared to the Nobel population results both from the inclusion of all the universities and *technische Hochschulen* in the area and the fact that not all of the institutions invited to nominate candidates for the prizes did so. See Paul Forman, J. L. Heilbron, and Spencer Weart, "Physics circa 1900: Personnel, funding and productivity of the academic establishments," *Historical Studies in the Physical Sciences* 5 (1975), whole issue.

[5] Christa Jungnickel and Russell McCormmach, *Intellectual mastery of nature*, vol. 1, *The torch of mathematics, 1800–1870*, pp. 202–213, and vol. 2, *The now mighty theoretical physicists, 1870–1925*, pp. 59–71, 184–192 (Chicago, 1986).

but the proportion of chemists holding posts at the *technische Hochschulen* was higher. This was the result of the higher number of separate positions in applied chemistry – chemical technology, electrochemistry, photochemistry, metallurgy – than in applied physics. At the universities, the division of chemistry into organic and inorganic was of long standing and was recognized through separate chairs. A further organizational differentiation occurred early in the twentieth century when physical chemistry was recognized as an independent subject. The first chair was instituted at the University of Vienna in 1902 and held by Rudolf Wegscheider for 30 some years. Prague followed suit in 1911 with Victor Rothmund as the chairholder. Both had received their training at German universities, Wegscheider at Berlin under Hans Landolt and Rothmund in Göttingen under Nernst.

A strict focus on the designation of positions, courses, and institutes overlooks the effects of language and nationality conflicts on the organization of higher education in the multinational empire. This vast topic can only be treated in barest outline, that is, for what it suggests about conditions specific to this periphery. These conditions were not really set by the authority of the emperor and the central bureaucracy, for the corporatist prerogatives of the universities in the matter of appointments, for instance, were of long standing. In this respect, the universities in the empire, especially during the time of its disintegration, probably had more autonomy than the German ones, which were formally under the control of the *Länder* but also had to suffer the interferences of the Reich bureaucracy.

More decisive were the ways culture and science became bound with the nationalist aspirations of the non-German language groups (in particular, Magyar and Czech) in the empire. In Hungary, Magyarization was so successful (except, of course, for the non-Magyar minorities) during the last 30 years of the Austro-Hungarian Empire that in the end both teaching and research were done in the national tongue. By contrast, in Bohemia and Moravia, the resistance of the German-speaking population to Czech demands for cultural autonomy produced a particularly virulent form of nationalism on the part of both groups. It was forcefully expressed by the persistent demands of the Czechs for their own institutions of higher education. Already in 1868, separate German and Czech *technische Hochschulen* were established. The chief symbol of national aspirations, however, was a separate Czech university, an aim realized in 1882, when Charles University was split in two. The doubling of chairs and other positions caused an unprecedented expansion of employment for scientists, but it may also have caused a temporary decline in the quality of teaching and research.[6]

Set in the mold of the German universities, the Austro-Hungarian ones did not have strong reasons to innovate. The two major innovations were the result of conditions, both human and natural, that were peculiar to the area. One was the joining of several strands of meteorology and geophysics into *Lehrkanzeln* for cosmic physics; the other, the creation of a unique research institution, the Institute for Radium Research (Institut für Radiumforschung) in Vienna. Because both had some bearing on nominations for the Nobel prizes, they will be described briefly here.

The special status given cosmic physics in Central European universities was mostly due to the organizational efforts of Julius Hann, who was director of the Central Office for Meteorology and Geodesy (Centralanstalt für Meterologie und Erdmagnetismus) and professor of physics at Vienna (1877–1897). It was Hann who pushed through the provision that made cosmic physics obligatory in the university exams for would-be teachers in the secondary schools. This provided the rationale in the universities for Hann's effort to join meteorology and geophysics into the entity known as cosmic physics. He was instrumental in the creation of separate *Lehrkanzeln* in cosmic physics, first, at the University of Graz, where one of his students, J. M. Pernter, held the chair (1890–1897), and then at Vienna, where Hann himself became the chairholder (1900–1910). Other students of Hann's who became professors of cosmic physics were R. Spitaler at the Prag German university, P. Czermak and W. Trabert (Trabert was the author of *Lehrbuch der Kosmischen Physik*, 1911) at Innsbruck, and V. Conrad at Czernowitz. The field failed to develop intellectual unity, however, probably because there were few common concepts and methods linking the very different phenomena of the earth and the atmosphere that were brought under the label of cosmic physics. After World War I, the *Lehrkanzeln* became

[6] Robert A. Kann, *The multinational empire: Nationalism and national reform in the Habsburg monarchy, 1848–1918*, 2 vols. (New York, 1950); and *Die Teilung der Prager Universität 1882 und die intellektuelle Desintegration in den böhmischen Ländern* (Munich, 1984), pp. 25–36 and 203–208.

merely institutional arrangements or were broken up into constituent parts such as meteorology and geophysics.[7]

The Institute for Radium Research in Vienna drew on a unique natural resource of the Austro-Hungarian Empire: pitchblende, mined in Joachimsthal, in Bohemia, from which radioactive substances were extracted. This was the source of the radium chloride prepared by the Curies in 1898. The research and medical uses to which this precious material was immediately put caused a steep rise in prices and a concern in the Vienna Academy of Sciences about meeting domestic research needs. The academy's Commission on Radioactive Substances, established in 1901 with Franz S. Exner as president, acted vigorously to set aside materials and was also able to lend some to foreign researchers, among others Ernest Rutherford and William Ramsay. Through the complex process of preparing radium chloride from pitchblende (10,000 kilograms yielded only 4 grams), which involved controlling intermediary and by-products, Vienna physicists were in on the ground floor of the burgeoning field of radioactivity. In the 10 years following the Curies' discovery, more than 50 publications came from Vienna alone.

The combination of technical expertise and access to research materials was institutionalized in 1908, when a private donor, Dr. Karl Kupelwieser, put up the funds to build and equip an institute. In 1910, the Institute for Radium Research of the Vienna Academy, the first of its kind, was inaugurated. Led by Stefan Meyer, the institute became an important center for the development of instrumentation in radiochemistry and atomic physics. Viktor Hess discovered cosmic radiation in 1912 using apparatus developed at the institute, and George de Hevesy, together with Fritz Paneth, carried out his first investigations of isotopes as tracers in chemical processes at the institute in 1912–1913. The institute was also engaged in testing standard units of radioactivity

---

[7] Ed. Brückner, "Julius Hann," in Akademie der Wissenschaften in Wien, *Almanach für das Jahr 1922* (Vienna, 1923), pp. 151–160; Gerhard Oberkofler, "Die Lehrkanzel für Kosmische Physik (1890–1955)," in *Die Fächer Mathematik, Physik und Chemie an der Philophischen Fakultäten Innsbruck bis 1945.* Veröffentlichungen der Universität Innsbruck 66 (Innsbruck, 1971), pp. 133–150. I am grateful to Lewis Pyenson (University of Montreal), who lent me his notes on cosmic physics during the Austro-Hungarian Empire and the outline of an article with the instructive title "When is an interdisciplinary phenomenon not a new discipline? A note on cosmic physics circa 1900." On the development of cosmic physics in Scandinavia, see Elisabeth Crawford, *The beginnings of the Nobel institution: The science prizes 1901–1915* (Cambridge and Paris, 1984), pp. 58–59.

for the International Radium Standard Commission of which Meyer was secretary.[8]

The developments described here, both those that followed the German model and those that diverged from it, made for a considerable degree of autonomy when it came to the area's ability to train and provide jobs for its scientists. A measure of the degree of educational self-sufficiency is the proportion of doctoral degrees granted by local institutions. Already in the 1880s, this was higher in the Austro-Hungarian Empire than in the United States, for instance, and it increased further with time. Of the 74 east-central European scientists for whom information is available, around 80 percent of those who received their doctorates before 1910 had earned their degrees at an institution in the empire. This was the case for 100 percent of those who received their doctorates after 1910.[9] Doctoral or postdoctoral training outside the region was most often done at German universities except for a few specialists in chemical technology, who trained in Switzerland.

Another measure of self-containment in regard to university positions is the number of individuals coming from the outside. Given the affinities of language and culture between Germany and the Austro-Hungarian Empire, a certain amount of "colonization" of the universities in the region by German imports would not have been surprising. There were only four members of the population, however, who were both born and educated outside the area, two of whom worked in technology. The inverse phenomenon of scientists leaving for lack of jobs and opportunities cannot be ascertained fully using the Nobel population because it counts as members only those who stayed in the area for a minimum of eight years. It may be significant, though, that of those who left before World War I or in the interwar period before the Anschluss, there were

---

[8] Institut für Radiumforschung, "Festschrift des Institutes für Radiumforschung anlässlich seines 40-jährigen Bestandes (1910–1950)," *Sitzungsberichte, Akademie der Wissenschaften, Wien, Mathematisch-Naturwissenschaftliche Klasse* 159 (1950): 1–57 (articles by, among others, S. Meyer, V. Hess, G. von Hevesy, and F. Paneth); Stefan Meyer, "Das erste Jahrzehnt des Wiener Instituts für Radiumforschung," *Jahrbuch der Radioaktivität und Elektronik* 17 (1920): 1–29; Roger Stuewer, "Artificial disintegration and the Cambridge-Vienna controversy," in Peter Achinstein, ed., *Observation, experiment and hypothesis in modern physical science* (Cambridge, 1985), pp. 239–307.

[9] The study of American candidates for the Nobel prizes in physics and chemistry (Chapter 6) shows that of those who earned their doctorates or entered the profession before World War I about 50 percent had doctorates from foreign, most often German, universities, 40 percent from American ones, and 10 percent had no doctorate.

many returnees, among them such prominent scientists as Ludwig Boltzmann, Viktor Hess, and Erwin Schrödinger.[10] A major exception were
the Hungarian physicists and chemists who emigrated early in their
careers, either for political reasons, or for lack of satisfactory conditions
for research, or both. Many of them – George de Hevesy, Theodor von
Karman, Leo Szilard, and Eugene Wigner, for instance – went on to
win Nobel prizes or to do prizeworthy work. They illustrate how the
periphery, given a particular set of historical, political, and intellectual
conditions, contributes to disciplinary development at the center through
the migration of talented scientists.[11]

The research orientations of candidates
for the Nobel prizes in physics and chemistry

The hierarchy of positions described earlier reflected different approaches to research: Students attended physics and chemistry laboratory courses as part of their training, postulants for academic jobs
carried out original investigations to qualify, and professors had come
to see research as part of their duties. Although this was the same division
of labor as in Germany, it produced research orientations that were not
simply copies of work carried out at the center. These orientations are
illustrated by the work of the 17 candidates for the Nobel prizes in
physics and chemistry between 1901 and 1939.

Nominations for the Nobel prizes are not unequivocal indications of
original contributions to knowledge because scientists are sometimes
nominated for reasons unrelated to such contributions. Taking into
account both the nature of the candidates' contributions and their support among the nominators yielded three categories that were labeled,

[10] Viktor Hess worked at the U.S. Radium Corporation in New Jersey from 1920 to 1923. He
left Austria permanently in 1938, when he was dismissed from his professorship on account of
his strict Roman Catholicism. Erwin Schrödinger spent his scientifically most productive years
at the universities of Zurich and Berlin before taking up a post at Oxford in 1933. He returned
to Graz 1936 to 1938 and, after the war years spent in Dublin, to Vienna in 1956. His stay at
Berlin University did not exceed the eight years that would have made him a German national
according to our definition of "working nationality," hence, he has been maintained as an
Austrian.
[11] Gabor Pallo, "A case from the periphery – The background of the 'Hungarian phenomenon,'"
paper presented at the International Congress of History of Science, 1985; Gabor Pallo, "Why
did George Hevesy leave Hungary?" *Periodica Polytechnica (Chemical Engineering)* 30 (1986): 97–
115; J. D. Cockcroft, "George de Hevesy, 1885–1966," *Biographical Memoirs of Fellows of the
Royal Society* 13 (1967): 125–166.

respectively, "international luminaries," "favorite sons," and "stay-at-home innovators." The following discussion gives examples of scientists belonging to each category.

The vigorous tradition in theoretical physics at Vienna and Graz produced the two chief *international luminaries* in the population: Ludwig Boltzmann and Erwin Schrödinger.[12] The contributions of both, Boltzmann in statistical mechanics and Schrödinger in quantum mechanics, advanced theoretical physics by addressing problems central to the discipline at the time. Their role in this respect was reflected in the broad international support of their candidacies. Boltzmann was nominated, among others, by Max Planck, and Schrödinger by Niels Bohr, Louis and Maurice de Broglie, and Einstein. To these two international luminaries, one may well add Ernst Mach, like Boltzmann a nonwinner, whose Vienna chair in the history and theory of science, especially physics, was instituted for him in 1895 in recognition of his international position in these fields. He was nominated, among others, by H. A. Lorentz.

Other than these three, there were lesser international luminaries, who took up problems treated by more restricted international specialty communities. Much of the support for their candidacies came from other members of those communities. Victor Hess's discovery of cosmic radiation in 1912 was an early contribution to a problem area that did not become a concern of physicists until the 1920s and 1930s. Through a series of daring balloon ascents, Hess established that the increased levels of atmospheric ionization observed at high altitudes were indeed of cosmic origin. His sharing the Nobel physics prize of 1936 with C. D. Anderson, whose much more recent work had confirmed that positrons make up the particles of cosmic rays, was a way for the Swedish Academy of Sciences to pay homage to one of the pioneers in the field.[13]

Another lesser light was Carl Auer von Welsbach. He was part of the small group of researchers in different countries who devoted their lives

---

[12] There was continuity between Boltzmann and Schrödinger in the person of Fritz Hasenöhrl, who was Boltzmann's favorite student and Schrödinger's teacher. Hasenöhrl succeeded Boltzmann in the chair of theoretical physics at Vienna but was killed in World War I. See Jungnickel and McCormmach, *Intellectual mastery of nature*, vol. 2, pp. 191–192.

[13] Berta Karlik and Erich Schmid, *Franz Serafin Exner und sein Kreis: Ein Beitrag zur Geschichte der Physik in Österreich* (Vienna, 1982), pp. 121–126; Yataro Sekido and Harry Elliot, eds., *Early history of cosmic ray studies: Personal reminiscences with old photographs*. Astrophysics and Space Science Library 118 (Dordrecht, 1985), pp. 3–31.

to the rare-earth elements. The great difficulties encountered in the separation of rare-earth elements made for much controversy in their isolation. In the first decade of the century, Auer von Welsbach was involved in a lengthy dispute with George Urbain at the Ecole des Mines in Paris concerning the separation of ytterbium into two new elements (neo-ytterbium and lutetium) and the naming of these. Auer von Welsbach was most widely known, however, for the invention of the Auer method of gas lighting using oxides that enhanced the luminosity of the flame following the principle of the Bunsen burner. The earnings from the manufacture and licensing of his invention in many countries made it possible for Auer to set himself up in a private laboratory. He thus became one of the few nonacademics in the population.[14]

The *favorite sons* were the direct opposites of the international luminaries. As the term indicates, their support came almost exclusively from within their own countries, if not their own universities. Their contributions were quite specialized and, although useful, were hardly of a kind to produce excitement. Among favorite sons with strong reputations in their respective national contexts were Joseph Maria Eder, who held the chair of photochemistry at the *technische Hochschule* in Vienna and was an innovator in the application of photometry to the study of spectra; Loránd von Eötvös, professor of physics at Budapest, who spent a lifetime perfecting the torsion balance, which provided high-accuracy measurements of the equality of gravitational and inert masses; and Hans Zdenko Skraup and Ernst Späth, professors of organic chemistry at Graz and Vienna, respectively. Together they represented two generations of work on alkaloids, particularly quinoline, which Skraup was the first to synthesize.[15]

The category of *stay-at-home innovators* included scientists who remained geographically peripheral to major centers of scientific activity. They did not leave the region, nor were they attracted to Vienna as the center of Austro-Hungarian science. Their choice of problems and the

---

[14] Jean D'Ans, "Carl Freiherr Auer von Welsbach," *Berichte der Deutschen Chemischen Gesellschaft* 5A (1931): 59–92.

[15] A biography of Eder is found in Josef Maria Eder, *History of photography* (New York, 1945); L. Marton, "Eötvös, Roland, Baron von," *Dictionary of scientific bibliography* (hereinafter referred to as *DSB*) 4 (1971): 377–381; Ernst Philippi "Hans Zdenko Skraup: Ein führender Chemiker organischer Richtung," in Fritz Knoll, ed., *Österreichische Naturforscher, Ärzte, und Tekniker* (Vienna, 1957), pp. 49–51; Friedrich Wessely, "Ernst Späth: Ein Meister der Erforschung organischer Naturstoffe," in Fritz Knoll, ed., *Österreichische Naturforscher*, pp. 55–57.

contributions they made were original in the sense that they did not evince a high degree of competitive emulation of the center. Julius Hann has already been mentioned for his introduction of cosmic physics into the universities of the Austro-Hungarian Empire at the turn of the century. His main contribution to meteorology was his studies of föhn – the strong, warm, down wind prevalent in the Alps – which he showed to be a climatological phenomenon specific to certain mountainous regions rather than originating, as was thought previously, in the Sahara.[16]

A similar independence from major trends at the center was shown after World War I by two chemists, Fritz Pregl and Jaroslav Heyrovsky. Pregl spent most of his career at Graz, where he held the chair of medical chemistry from 1913 to 1930. There he developed the basic techniques of microanalysis of organic materials that proved essential to the growth of biochemistry where the large test samples required for more traditional analytical methods were not available. Heyrovsky trained in electrochemistry under F. G. Donnan and received both his bachelor of science and doctoral degrees from University College, London. He returned to Prague University after World War I and remained there as professor of physical chemistry until 1959. Having developed polarography in the early 1920s, he spent the next 40 some years perfecting what became the standard analytical technique for identifying and determining the concentration of ions by electrolysis of a solution. The main tool was the polarograph, a simple instrument built in Heyrovsky's laboratory. Both Pregl and Heyrovsky were leaders of research schools that attracted foreign students to their respective universities. Their international reputation was demonstrated by the support given their candidacies. Both men eventually earned Nobel prizes in chemistry, Pregl in 1923 and Heyrovsky in 1959.[17]

The differences in the work orientations of the three types of peripheral scientists are reflected in their publication profiles. Overall, all three types had a high lifetime productivity, a mean of more than 100

---

[16] Brückner, "Julius Hann," pp. 151–160.
[17] Mikulas Teich, "Heyrovsky, Jaroslav," *DSB* 6 (1972): 370–376; J. A. V. Butler and P. Zuman, "Jaroslav Heyrovsky, 1890–1967," *Biographical Memoirs of Fellows of the Royal Society* 13 (1967): 167–191; P. Zuman, "With the drop of mercury to the Nobel Prize," in John T. Stock and Mary Virginia Orna, eds., *Electrochemistry, past and present.* ACS Symposium Series 390 (Washington, D.C., 1989), pp. 339–369; and Anton Holasek, "Fritz Pregl," in Kurt Freisitzer, Walter Höflechner, Hans-Ludwig Holzer, and Wolfgang Mantl, eds., *Tradition und Herausforderung: 400 Jahre Universität Graz* (Graz, 1985), pp. 257–264.

Critical and empirical studies

Table 5. *Publications in journals within and outside east-central Europe by category of peripheral scientist*

	Publications			
	Local (*N*)	Foreign (*N*)	Total (*N*)	% foreign
International luminaries	249	620	869	71
Favorite sons	332	188	520	28
Stay-at-home innovators	470	140	610	23

Table 6. *Publications in journals within and outside east-central Europe: international luminaries*

		Publications	
	Dates of first and last publication	Local (*N*)	Foreign (*N*)
Carl Auer von Welsbach	1883–1926	16	16
Ludwig Boltzmann	1865–1905	53	90
Felix Ehrenhaft	1902–1951	25	99
Viktor Hess	1905–1964	49	103
Ernst Mach	1859–1916	92	117
Erwin Schrödinger	1910–1966	14	195
Total		249	620

Percentage foreign publications of total: 71.
Productivity mean for group: 144.

journal publications. In this respect, they abided by Michael Faraday's precept "Work, Finish, Publish!" That there were important differences between the three types, however, is revealed by the data in Tables 5, 6, 7, and 8. In Table 5, the total number of publications in journals in the region (labeled "local") as compared with those outside it ("foreign") is shown for each type; in Tables 6, 7, and 8 the profiles of individual scientists are displayed. As could be expected, the international luminaries had the largest percentage of publications in journals outside east-central Europe (71 percent) and the lowest (29 percent) in local journals. The inverse relationship of around 70 percent local publica-

Table 7. *Publications in journals within and outside east-central Europe: favorite sons*

	Dates of first and last publication	Publication	
		Local (N)	Foreign (N)
Joseph Maria Eder	1877–1931	120	59
Friedrich Emich	1883–1933	38	26
Loránd von Eötvös	1869–1922	82	7
Hans Zdenko Skraup	1874–1911	92	19
Siegmund Strauss	1923–1937	0	18
Total		332	129

Percentage foreign publications of total: 28.
Productivity mean for group: 92.

Table 8. *Publications in journals within and outside east-central Europe: stay-at-home innovators*

	Dates of first and last publication	Publication	
		Local (N)	Foreign (N)
Julius von Hann	1863–1921	261	6
Jaroslav Heyrovsky	1921–1959	90	64
Fritz Pregl	1898–1930	6	21
Emil Votocek	1895–1930	113	49
Total		470	140

Percentage foreign publications of total: 23.
Productivity mean for group: 152.

tions and less than 30 percent foreign ones obtains for the favorite sons. Somewhat surprisingly, the stay-at-home innovators had the lowest portion of publications in foreign journals (23 percent), while at the same time being the most productive group (their mean lifetime productivity was 152 articles as compared with 144 for the international luminaries).

That the stay-at-home innovators scored so low on foreign publication was probably due to the strategies they employed to publish the results

Critical and empirical studies

of their own research and that of their students. The four stay-at-home innovators were all editors or co-editors of journals published in east-central Europe: Hann of the *Meteorologische Zeitschrift*, a joint publication of the German and Austrian meteorological societies, Pregl of *Mikrochemie*, and Heyrovsky and Votocek of *Czechoslovak Chemical Communications*. The last-mentioned is the best example of a conscious attempt to overcome the isolation that a geographically peripheral location and a minority language could impose.

In 1928, Heyrovsky took the initiative to prepare a statistical study of the publication patterns of Czech chemists during the first decade of the republic.[18] His survey showed that the habit of publishing the same article in a "local" journal and a "foreign" one was widespread. Thus, while close to half of the output was published in foreign journals, often these articles had already appeared in Czechoslovakia. In this way, Czech chemists could keep their linguistic distinctiveness, and also promote science for their newly created nation while not losing the opportunity to make their work known abroad.[19] Rather than scatter the foreign output over innumerable journals, Heyrovsky proposed that it be regrouped in a Czech journal published in English and French, a proposal that led to the creation of *Czechoslovak Chemical Communications*, which became a widely disseminated journal.

Recognition at the center

Restricted to a relatively small group of well-known, highly productive physicists and chemists, the publication patterns of the candidates for the Nobel prizes are not typical either of scientists in the region or of scientists in general.[20] A more serious restriction of the data presented

[18] "Přehled zahraničních publikací československých chemiků v prvním desítiletí republiky" [Publications of Czech chemists in foreign journals during the first decade of the Czechoslovak Republic], *Chemické listy* 22 (1928): 413–415.
[19] The strategy of publishing the same article, locally, in the native language, and, internationally, in German, French, or English, was employed by scientists in Finland, Norway, and Sweden at the turn of the century.
[20] There is no doubt that the candidates were much more productive than their peers. A rough calculation puts their mean number of papers per year during a lifetime of scientific work at around 3.5. This was the same as for the high scorers described by de Solla Price in *Little science, big science*. In Price's terms, "about one-third of the literature and less than one-tenth of the men are associated with high scores." Derek J. de Solla Price, *Little science, big science* (New York, 1963), p. 49.

in the tables, however, is that they do not show whether or not the considerable number of articles produced by this group of peripheral scientists attracted any attention from the center. These questions are addressed drawing on a major resource for the historical study of publication and citation patterns, the *Physics citation index, 1920–1929 (PCI)* compiled by Henry Small.[21] The general organization and utility of the *PCI* have been described both by Small himself and by Lewis Pyenson and M. Singh.[22]

All members of the Nobel population whose age indicated that they were active in research in the decade 1920 to 1929, and hence could have published, were identified. The *PCI* is constituted around authors cited in 16 rather loosely defined core "physics" journals, and hence, they include both chemists, especially physical chemists, and mathematicians. None of the core journals was published in east-central Europe, which makes the citation counts truly representative of the recognition that authors from the area received outside the region. Most of the core journals were published in major science-producing countries, Germany and England, in particular.[23] This exercise yielded a list of 20 authors who had either published or been cited in one of the core journals (see Tables 9 and 10).

The publications were counted separately, depending on whether they had appeared in one of the 16 core journals or in other journals either in the region or outside. For the core journals, the *PCI* was used; for other journals, the *Biographisch-Literarisches Handwörterbuch zur Geschichte der exacten Wissenschaften* compiled by J. C. Poggendorff. This exercise also served to identify articles by the several authors in the source index who appeared as different people despite their being the same individual (for example, Schrodinger and Schroedinger). The authors were then ranked according to the percentage of core articles of

[21] Henry Small, *Physics citation index, 1920–1929*, 2 vols. (Philadelphia, 1981).
[22] Henry Small, "Recapturing physics in the 1920's through citation analysis," *Czechoslovak Journal of Physics* 36B (1986): 142–147; and Lewis Pyenson and M. Singh, "Physics on the periphery: A world survey," *Scientometrics* 6 (1984): 280–282.
[23] The following journals were used to construct the *PCI*: Germany – *Zeitschrift für Physik, Annalen der Physik, Physikalische Zeitschrift*; England – *Monthly Notices of the Royal Astronomical Society, Philosophical Magazine, Philosophical Transactions of the Royal Society of London, Section A, Proceedings of the Royal Society of London, Section A, Proceedings of the Physical Society*; United States – *Astrophysical Journal, Physical Review*; France – *Annales de physique, Journal de physique*; Italy – *Il Nuovo Cimento*; Denmark – *Matematisk-Fysiske Meddelelser*; Netherlands – *Physica*; Japan – *Proceedings of the Physico-Mathematical Society of Japan*.

Critical and empirical studies

the total number of articles. Table 9 gives the rank order of the 19 authors who published in core journals.

References to articles by the 20 authors who were cited in core journals were tabulated using the *PCI*. The index is limited to references made in articles appearing in the core journals, 1920 to 1929, but they may concern any work by the cited author: books, manuals, undated publications, articles appearing before 1920, and so forth. Citations by the authors to their own work were excluded. In order to have a homogenous assemblage of citations, only those made to articles appearing between 1920 and 1929 were tabulated. Altogether 630 references were made to articles by the 20 authors on the list. Table 10 ranks the cited authors according to the percentage of citations received.

The conditions that governed the visibility at the center of peripheral physicists are apparent from a comparison between Tables 9 and 10. With the exception of three individuals (Heinrich Mache, Gustav Hüttig, and Eduard Haschek), the names on the two lists are identical. As a rule, then, publishing in core journals was a condition for being cited in those journals. The ineluctable nature of this condition is brought out further when one examines the citation records of nine physicists who published in the 1920s but not in core journals. Only two of these (Haschek and Hüttig) figured in the citation index and were therefore included in Table 10.

The comparison can be taken one step further by looking at the ranks of the authors in Tables 9 and 10. With a few exceptions, the first ten authors formed a group that was the same for both lists, accounting for 73 percent of the articles in core journals and 80 percent of the citations. A strong showing in core journals thus went hand-in-hand with high citation visibility. There was further symmetry in that the largest proportion of both the articles by east-central European authors and the references made to work by these authors (which, it should be recalled, could concern articles in any journal) appeared in core journals published in Germany (89 percent of the articles and 73 percent of the references). "Publish or perish" on the southern periphery of Central Europe in the 1920s, then, meant getting attention for one's work in Germany.

The names on the lists in Tables 9 and 10, as well as their rank, hold some surprises. That Schrödinger should hold the lion's share of both publications and citations in core journals is hardly surprising, for his four-part series on wave mechanics, "Quantisierung als Eigenwert-

Center-periphery relations in science

Table 9. *Ranking of selected members of the Nobel population in east-central Europe by number of publications in core journals, 1920–1929*

	Publications Core (N)	Other (N)	Total (N)	Share of core (%)
1. *Erwin Schrödinger	27	14	41	18
2. Reinhold Fürth	17	24	41	11
3. Karl Lichtenecker	15	8	23	10
4. Heinrich Rausch von Traubenberg	12	11	23	8
5. *Hans Thirring	11	15	26	7
6. *Felix Ehrenhaft	10	5	15	7
7. *Viktor Hess	10	17	27	7
8. Friedrich Kottler	9	7	16	6
9. *K. W. F. Kohlrausch	7	15	22	5
10. *Hans Benndorf	6	15	21	4
11. *Karl Przibram	4	26	30	3
12. Jaroslav Heyrovsky	4	32	36	3
13. *Ludwig Flamm	4	3	7	3
14. *Stefan Meyer	3	18	21	2
15. *Heinrich Mache	3	13	16	2
16. *Gustav Jäger	2	3	5	1
17. Václav Posejpal	2	27	29	1
18. *Egon von Schweidler	1	8	9	1
19. *Franz Aigner	1	3	4	1
Total	148	264	412	100

Note: Members of the Exner circle are indicated with an asterisk.
Sources: Henry Small, *Physics citation index, 1920–1929* (Philadelphia, 1981); and J. C. Poggendorff, *Biographisch-Literarisches Handwörterbuch zur Geschichte der exacten Wissenschaften* (Leipzig, 1863–).

problem," published in *Annalen der Physik* in 1926, received instant recognition. (He was awarded the Nobel prize, together with Paul Dirac, in 1933.) That the other candidates and eventual prizewinners, Hess and Heyrovsky, do not rank high either on publications or on citations in core journals can be attributed to their being at stages in their careers that did not yield large numbers of significant publications. Hess did applied work in radioactivity in the United States in the early 1920s and Heyrovsky's polarographic work had only just begun.

Critical and empirical studies

Table 10. *Ranking of members of the Nobel population in east-central Europe by number of citations in core journals, 1920–1929*

	Citations (N)	Percent
1. *Erwin Schrödinger	228	36
2. Reinhold Fürth	57	9
3. *Felix Ehrenhaft	50	8
4. Heinrich Rausch von Traubenberg	49	8
5. *K. W. F. Kohlrausch	35	5
6. *Viktor Hess	34	5
7. *Hans Thirring	30	5
8. Friedrich Kottler	28	4
9. *Karl Przibram	27	4
10. *Stefan Meyer	17	3
11. Gustav Hüttig	15	2
12. Karl Lichtenecker	15	2
13. *Hans Benndorf	11	2
14. Jaroslav Heyrovsky	11	2
15. *Ludwig Flamm	5	1
16. *Egon von Schweidler	5	1
17. *Eduard Haschek	4	1
18. *Gustav Jäger	4	1
19. Václav Posejpal	3	0.5
20. *Franz Aigner	2	0.5
Total	630	100

Note: Members of the Exner circle are indicated with an asterisk.
Source: Henry Small, *Physics citation index, 1920–1929* (Philadelphia, 1981).

Following Schrödinger in the rankings, there were three theoretical physicists – Reinhold Fürth, Karl Lichtenecker, and Heinrich Rausch von Traubenberg – all holding professorships at Prague University. Throughout the 1920s, they produced an impressive number of articles in core German journals, in particular, *Zeitschrift für Physik* and *Annalen der Physik*, that received a fair share of citations. Many of their papers seem to fall in the category that Small refers to as "theories of phenomena frequently encountered in the laboratory...in short, useful methods, procedures or formulas."[24] They also confirm the existence

[24] Small, "Recapturing physics in the 1920's," p. 146.

of a research milieu and tradition in theoretical physics, particularly electrodynamics, at Prague; previously the chair and institute there had been known mainly through Einstein's sojourn in 1911–1912.[25]

Among the remaining names on the lists in Tables 9 and 10, more than half had received their doctorates under Franz S. Exner at Vienna between about 1880 and 1910 and were considered members of the Exner circle. This group deserves closer examination since it points up the periphery's potential for innovative approaches, both intellectually and organizationally.

The Exner circle

Exner and his students at Vienna – Hans Benndorf, Stefan Meyer, Viktor Hess, and Erwin Schrödinger, to mention only a few – who came to dominate east-central European and in some respects international physics in the interwar period, did not form a research school but, more loosely, a circle. Opinions of what constitutes a research school vary. Generally, however, a research school has been seen as involving a scientific leader who mobilizes resources in personnel and monies around a program in which a distinctive approach is brought to bear on a few sharply delimited problems within a broader field.[26] Exner's circle differed from this, most importantly perhaps, by the wide-ranging research interests of Exner himself: atmospheric electricity, radioactivity, color theory. He trained students by working with them on these problems, first in the old Institute of Physics at Türkenstrasse and, after 1913, in the new one at Boltzmanngasse, in seminars, and in informal discussion groups. Once they had received their *Habilitation*, his students moved out of his orbit, another difference with the conventional view of a research school, often to positions at other universities in the Austro-Hungarian Empire. Many of them continued to work in Exner's areas of interest. With time, common interests gave rise to collaborations between former Exner students that did not involve the "master" – yet another difference with the notion of a research school – but still ensured the continuing existence of the circle.

Exner's organizational talents were as far-reaching as were his re-

[25] Jungnickel and McCormmach, *Intellectual mastery of nature*, vol. 2, pp. 297–298.
[26] Gerald Geison, "Scientific change, emerging specialties, and research schools," *History of Science* 19 (1981): 20–40.

search interests, another aspect of his personality that made for a circle rather than a school. Once launched, his organizational schemes were most often carried forward by his students to whom he was only too happy to delegate responsibility. He was instrumental in the creation of the Institute for Radium Research, having presided over the committee of the Vienna Academy of Sciences that conceived the idea of the institute, and he also supervised its construction. Named director of the institute when it opened in 1910, he gave Stefan Meyer the lion's share of responsibility for running it. Meyer, in turn, made room at the institute for other members of the circle, among them, Egon von Schweidler, Eduard Haschek, Viktor Hess, and K. W. F. Kohlrausch. Exner moved on to other organizational tasks, in particular, planning and supervising the construction of the new Institute for Experimental Physics at the University of Vienna, which opened in 1913, and coordinating studies of atmospheric electricity carried out at a string of observation points in Austria.[27]

On his retirement in 1920, Exner left his students with a legacy of well-functioning structures for research that put them in good position to rank high on the lists of authors and citations in core journals throughout the 1920s. Of the ten top-ranking individuals on both lists, six and seven, respectively, were members of the Exner circle. Maybe because they shared the masters far-reaching research interests, Exner's students were able to produce survey articles for physics handbooks, which invariably led to a high citation count. For example, Hans Benndorf wrote the article on atmospheric electricity for the *Handbuch der Experimentalphysik* (1928); K. W. F. Kohlrausch the one on radioactivity in the same volume and also one on atmospheric electricity (together with Egon von Schweidler) in Graetz's *Handbuch der Elektrizität und des Magnetismus* (1918–1928); and Erwin Schrödinger the one on specific heats for the *Handbuch der Physik* (1926).

The presence of so many members of the Exner circle among the high scorers was not due solely, however, to strategic positions in their respective fields. "Scientific style" is notoriously difficult to substantiate. Still, Exner and his close collaborators showed a similarity of approach to their work that fits this notion and that may have made them more productive than their peers. One thing they had in common was the

---

[27] Karlik and Schmid, *Franz Serafin Exner und sein Kreis*, passim.

ease with which they moved from the microlevel to the macrolevel in physics. This may have been a consequence of their early training in atmospheric electricity, a microphenomenon measured under natural conditions and therefore susceptible to local variations (meteorological, geological, or other) on the macrolevel. Some of these data were gathered on scientific expeditions – Kohlrausch in the West Indies (1907–1908) and Mache in the Far East (1900–1901), for instance – which were also occasions for more traditional work in geophysics. Most of the members of the circle, in fact, had an interest in geophysics; Benndorf, for example, won renown for his work in seismology, and Hess and Mache, among others, applied their skills in radioactivity measurements to geology.[28]

Another characteristic that set Exner's students apart from their colleagues in Germany, for example, be they theoretical or experimental physicists, was their unusual skill in developing instruments and techniques to investigate the most varied phenomena. The prime example is Hess's adaptation of instruments for measuring radioactivity and atmospheric electricity to establish the existence of a hitherto unknown radiation in the upper atmosphere that could not result from radioactive emanations on earth and therefore had to be cosmic in origin.[29] In this case superior skills in instrumentation brought forth an important discovery. A too heavy reliance on instrumentation, however, sometimes made Exner and his students lose sight either of the original problem they had set for their research, or the theory behind it, or both. For instance, when Exner and Eduard Haschek wanted to use spectroscopy to study meteorites, they saw the need to bring improvements to spectrograms using new photographic techniques. Over 10 years, they defined some 100,000 spectral lines more accurately than before but never got around to the meteorites.[30]

An overreliance on superior skills when it came to instrumentation and measurement may have been the underlying reason for the Cambridge-Vienna controversy that dominated research on the artificial disintegration of the chemical elements between 1919 and 1927. "Cambridge" in this instance was the Cavendish Laboratory where Ernest Rutherford and James Chadwick plugged away at Rutherford's

---

[28] Ibid., pp. 101–141.
[29] Ibid., pp. 124–126.
[30] Ibid., pp. 74–75.

1919 discovery that the nucleus of nitrogen could be artificially disintegrated by bombarding it with radium chloride alpha particles. "Vienna" was the Institute for Radium Research where Hans Pettersson and Gerhard Kirsch challenged the Cambridge results by extending the range of disintegrable elements to the lighter elements (beryllium, magnesium, and silicon) found "nonactive" by Rutherford and Chadwick. The experimental results that were at issue in the controversy also gave rise to essentially different theoretical interpretations; Rutherford's satellite model of the nucleus of the atom was pitted against Pettersson's explosion model.[31] Pettersson, a Swedish guest researcher, had grown up in the circle of his father, Otto Pettersson, a chemist turned hydrographer, and Svante Arrhenius, whose approaches to physics were as wide-ranging as those of the Exner circle.[32]

To this background, Pettersson added an ingenuity in the design and construction of experimental apparatus that put him on a par with his colleagues at the Vienna Institute. This may have made him overconfident in experimental results that, in the final analysis, depended on human factors, principally the power of the human eye to detect scintillations from particles of different range. As it turned out, human fallibility could explain why the result of the Vienna group differed from that of the Cambridge group. Giving in to higher authority, Pettersson and his co-workers quietly dropped their investigations in 1927.[33] Artificial disintegration had shown itself to be too hard a nut to crack for the Vienna group, but this was only very indirectly the result of its peripherality.

Conclusion and discussion

We are now in a better position to assess in what respects physics and chemistry in east-central Europe during the first three decades of the twentieth century can be held to have been peripheral in the sense given the term by Ben-David. Peripherality in this sense, it should be recalled, refers to the inability to compete with the center for reasons of small size of the scientific enterprise, linguistic isolation, or physical location. These conditions all applied in varying degrees to physics and chemistry

---

[31] Stuewer, "Artificial disintegration and the Cambridge-Vienna controversy," pp. 239–307.
[32] Crawford, *The beginnings of the Nobel institution,* pp. 54–59.
[33] Stuewer, "Artificial disintegration and the Cambridge-Vienna controversy," pp. 284–294.

in east-central Europe. According to Ben-David's model, these conditions should have made this particular periphery dependent on its German center because they impeded the organizational innovations, which alone determine the scale and orientation of scientific activities.

Indisputably, there were two areas in which east-central Europe assumed the role of periphery with respect to the German center:

1. *Citation visibility.* East-central European physicists, even the very productive ones, were clearly at a disadvantage when it came to being cited in the core physics journals, mainly German, unless they themselves published in these journals.

2. *Migration patterns.* For the east-central European scientists who migrated, the center – initially Germany and then the United States – was a powerful pole of attraction. This condition applied regardless of whether the migration was the result of the push out of the homeland, as in the case of some Hungarian scientists, who left on account of the political upheavals following World War I, or the pull that the center with its superior facilities for research exerted on other peripheral scientists.[34]

In most other respects, however, east-central Europe did not constitute a periphery in the sense given the term by Ben-David. True, with a few exceptions, the region's physicists and chemists did not compete with their colleagues at the center, but neither did they slavishly imitate the work orientations at the center – this despite the fact that the institutional arrangements for teaching and research at the universities were modeled on those of Germany. To some extent, the institutional setup also produced similar scientific developments, but this only with respect to very broad and general features, such as the differentiation of the disciplines of physics and chemistry into such independent specialties as theoretical and mathematical physics and physical chemistry.

The area did not conform to the pattern suggested by Ben-David, however, in that it was the site of organizational and scientific innovations that were largely independent of developments at the center. Some of

---

[34] It is curious that Ben-David should have stressed (*Scientist's role in society*, p. 172) the difficulties that scientists from small nations speaking minority languages faced when migrating to the center, as he himself was a most successful migrant from Hungary to Israel and then to the United States. (Gad Freudenthal and J. L. Heilbron, "Eloge: Joseph Ben-David, 1915–1986," *Isis* 80 [1989]: 659–663.)

these are revealed by the typology of candidates for the Nobel prizes between 1901 and 1939 presented earlier. Although the international luminaries and favorite sons reflected the traditional view of center-periphery relations, the stay-at-home innovators did not – this because of the strategies they employed to overcome the disadvantages of a geographically peripheral location and, in the case of the Czechs, the legacies of a militant nationalism that made them linguistically peripheral as well. They worked on narrowly defined problems, polarography in the case of Heyrovsky and microanalysis in that of Pregl, for instance, where innovation in instrumentation and technique provided a definite advantage; they founded their own journals – *Mikrochemie* and *Czecho-slovak Chemical Communications* – so as to hold the center's monopoly on communications in check; and they profited from the lack of competition for students and resources by establishing strong research schools. It is particularly interesting that neither Heyrovsky nor Pregl considered Vienna the scientific center of the region. Instead, they sought and found support for their schools in their respective localities and abroad.

As shown by the description of the Exner circle, a high profile of publication and citation in core journals did not have to result from a mere copying of predominant orientations at the center. In this case, it was the consequence, rather, of structures for research and a research style that could hardly have developed at the center. The longevity of the circle can be attributed to its being protected from the competition that in Germany would inevitably have led to the mobility and dispersion of its members. In the present case, with the exception of Schrödinger, all the members of the circle remained at universities in the region throughout their careers. The wide-ranging approach to research – embracing at the same time theoretical physics, atomic physics, and geophysics – that characterized members of the circle was more a function of the small scale of the scientific enterprise, though, than of distance from the center.[35] An eclectic approach to research was the price that scientists in the region had to pay for their institutional autonomy. It did not mean isolation and lack of recognition at the center, however,

[35] This point has been made with respect to present-day Finnish science. See Erkki Kaukonen and Pirkko-Liisa Rauhala, "Structural transformation of science: Trends and problems of scientific integration," *Science studies* 2 (1989): 26–27.

but made room for an integrative approach, illustrated by the many articles that Exner's students published in various physics handbooks.

To the extent that the region constituted a periphery within Central Europe in more than a geographic sense, then, this was relative and variable depending on the particular aspect of scientific activity examined. Rather than being separate entities, Germany and such outlying areas as east-central Europe or Scandinavia resembled the jagged pieces of a puzzle in the making where close proximity on some counts coexisted with distance on others. To get a picture of such a multifaceted reality, the center-periphery dichotomy might profitably be replaced by a notion such as "effective professional distance." This could be evaluated, and even measured, in different localities, within the same country or in different ones, along the several dimensions that constitute scientific activity: training of students, experimental research, formal and informal communication, and so on. Drawing such a picture would no doubt reveal the complementarity of much scientific activity carried out in different countries, an aspect that has been obfuscated by the emphasis on competition in Ben-David's work.

~~~~~~

5

National purpose and international symbols: the Kaiser-Wilhelm Society and the Nobel institution

The Kaiser-Wilhelm Society for the Advancement of Science (known as KWG, which in German stands for Kaiser-Wilhelm-Gesellschaft zur Förderung der Wissenschaften) and the Nobel institution had in common the goal of promoting elite science. To accomplish this purpose each had, a few years after its establishment, about the same income and capital including buildings.[1] That what each bought with these monies was altogether dissimilar was a consequence of the very different purposes of the two institutions. For this reason, their interactions, which concerned important aspects of the workings of both, are particularly instructive.

The purpose of the KWG, founded in 1911, was first and foremost to tap private money by recruiting paying members and other donors to the society and only secondarily to use government assistance to build, equip, and operate independent research institutes in different branches of science. These were the Kaiser Wilhelm Institutes (KWIs). It was a national purpose in that the stated rationale for the enterprise was that the productivity of professors (the main research force in Germany) had been curtailed by their obligation to teach an ever-growing number of students, who, to make matters worse, made important demands on equipment and assistants. It is not the purpose here to determine how well these claims fit the facts. But they would not have carried conviction

---

[1] Bernhard vom Brocke, "Die Kaiser-Wilhelm-Gesellschaft im Kaiserreich," in Rudolf Vierhaus and Bernhard vom Brocke, eds., *Forschung im Spannungsfeld von Politik und Gesellschaft: Geschichte und Struktur der Kaiser-Wilhelm/Max-Planck-Gesellschaft* (Stuttgart, 1990), pp. 145–151 and 155–157; and Elisabeth Crawford, *The beginnings of the Nobel institution: The science prizes 1901–1915* (Cambridge and Paris, 1984), p. 27.

had they not been coupled with the warning that, without separate research institutes, plentiful assistance, and whatever equipment they required, the German professoriate, which had dominated international research from its strongholds in university institutes, would be beaten by the British and the Americans, who had invented the privately endowed, independently directed research establishment: the Royal Institution, the Carnegie Institution, and the Rockefeller Institute, the major such organizations functioning in the first decade of the twentieth century.[2]

In a sense, the directorships of the KWIs might be considered prizes, because they went to senior men distinguished in research, and brought both research subsidies and enviable working conditions. But even considered as prizes, the KWI directorships differed in orientation from the Nobel awards. Though offered on the basis of performance, they were intended to enable future, not to commemorate past, research. Accordingly, they differed as much from Nobel prizes in their conception as the distinct national purpose of the KWG differed from the prized cosmopolitanism of the Nobel endeavor.[3]

The Nobel prizes were the antidote to these developments, for Nobel had conceived them as an aid to scholars, preferably younger ones, working alone in the heroic, nineteenth-century mold of a Pasteur or a Mendel. An important factor militating against such use of the awards, however, was their amount – about four times the annual salary of a full professor when they were first awarded in 1901. As a result, they came to be awarded to well-established scholars of international repute, exactly the kind of person who would be considered for a KWI directorship. University professors constituted a majority among the prize-winners and most professors then identified science with pure or basic knowledge. Hence, from the beginning there was a tension between Nobel's will, which specified "discovery" or "invention" (the latter term presumably recommending technology) as objects of reward, and the professional ideals of the professors and academicians who administered it. The same tension characterized, and even drove, the KWG.[4]

The Nobel institutes constituted yet another common bond between the KWG and the Nobel institution. The idea to use Nobel monies to

[2] vom Brocke, "Die Kaiser-Wilhelm-Gesellschaft im Kaiserreich," pp. 126–130.
[3] See Chapter 2, the section on "Internationalism and nationalism in the context of the Nobel institution."
[4] Crawford, *The beginnings of the Nobel institution*, chapters 3 and 6.

mount and maintain research laboratories in experimental science was launched by representatives of the scientific institutions named in the will (the Royal Academy of Sciences and the Karolinska Institute) in the course of the lengthy negotiations (1896–1900) over the statutes of the Nobel Foundation. They felt that the fortune of about 30 million crowns ($8 million) left by Nobel was so large that one-fifth of the interest on it seemed excessive for a prize, and especially a prize functioning as a stipend. The endowment for the Nobel institutes, which at the outset amounted to 300,000 crowns ($70,000) for each institute, was thus wrested from the prize money. Each prize awarder would decide whether or not to set up one or several institutes in its prize domain and also manage them.[5]

When the KWG was first being considered, in 1907–1909, only one Nobel institute had been set up, the Nobel Institute for Physical Chemistry, created for Svante Arrhenius at the Royal Swedish Academy of Sciences in 1904 and housed in temporary quarters until 1909. In the early plans for the KWIs in basic science, it appears as one of several examples to support the charge that Germany lagged behind foreign countries in the creation of independent research institutes. The single realization of the Nobel institutes and the grandiose plans drawn up for the KWIs make incongruous the references to Arrhenius's institute as a model for the KWIs.[6] Incongruous, but not poor propaganda: Assimilation with the Nobel institute invested the KWIs, before their birth, with the implied objective of furthering elite science, and insinuated that to compete with Britain, France, and the United States, Germany would have to mobilize its Alfred Nobels. Once created, however, the scale and functioning of the KWIs were so different from the Nobel institutes as to obviate further comparisons between the two.

The most important KWIs for basic science in terms of the size of their staffs and research facilities were established in 1911–1912.[7] The main basis of interaction between them and the Nobel institution was the prize. Three types of relationships were involved here: winning, being nominated, and nominating. Each may serve as a probe into the

---

[5] Ibid., pp. 72–76; and Elisabeth Crawford, "The benefits of the Nobel prizes," in Tore Frängsmyr, ed., *Science in Sweden: The Royal Swedish Academy of Sciences, 1739–1989* (Canton, Mass., 1989), pp. 235–240.
[6] vom Brocke, "Die Kaiser-Wilhelm-Gesellschaft im Kaiserreich," p. 126.
[7] Ibid., pp. 145–148.

workings of the KWG and its basic science institutes. Seven of these institutes will be considered: Chemistry, founded in 1911 (KWI für Chemie); Physical Chemistry and Electrochemistry, also 1911 (KWI für physikalische Chemie und Elektrochemie); Biology, 1912 (KWI für Biologie); Experimental Therapy, 1912 (KWI für experimentelle Therapie), renamed Biochemistry in 1936 (KWI für Biochemie); Physics, 1914 (KWI für physikalische Forschung); Aerodynamical Testing Station, 1918 (Aerodynamische Versuchsanstalt der KWG), reorganized in 1925 as Aerodynamic Research (KWI für Strömungsforschung); and Cell Physiology, 1931 (KWI für Zellphysiologie).

Interactions centered on the Nobel prizes

The interaction between the KWG and its institutes and the Nobel institution centered on the participation of KWG and KWI personnel in the process of prize selection. "Personnel" has been defined broadly to include not only the scientific staffs of the KWIs but also individuals serving in an administrative or advisorial capacity. The KWG, it should be recalled, was set up to enlist the support of wealthy individuals for the Society. This was accomplished through dues-paying members (there were 150 in 1911 and about 500 by the mid-1920s), who were expected to be altruistic, inasmuch as they received little in return for their 20,000 marks ($5,000) entrance fee and the 1,000 mark ($250) annual dues.

The 20 members of the governing body of the KWG, the Senate, were much more richly rewarded; they wore a specially designed dark-green velvet robe with the sleeves and collar in orange silk, and their meetings often opened or closed with a reception or some other ceremony attended by the kaiser. They returned the honors by an effusion of beneficence, contributing more than 4.5 million marks ($1.1 million) to the KWG between 1911 and 1914. They could well afford it, for among the senators were the two wealthiest men in Prussia – Gustav Krupp von Bohlen und Halbach and Guido Graf Henckel Fürst von Donnersmarck – as well as the owners of Berlin private banks and electrical and chemical industrialists (as Wilhelm von Siemens). The three scientist members of the Senate (Emil Fischer, J. H. van't Hoff, and Paul Ehrlich) were all Nobel prizewinners. The Senate elected the executive committee (*Verwaltungsausschuss*), whose president, together with the secretary-general, had responsibility for the day-to-day running of the society.[8] In addition to these central administrative and decision-making bodies, there was an advisory board (*Kuratorium*) for each KWI, made up of prominent scientists, representatives of ministries and scientific societies such as the Prussian Academy of Sciences, and local authorities.

In the following, the roles of prizewinners, nominees for the prize, and nominators will be examined in turn for the period 1911 to 1939. Because these roles had their rough correlates in the overall structure of the KWG and the KWIs, each points up some specific aspect of the workings of these organizations.

Prizewinners

Table 11 presents prizewinners in physics, chemistry, and physiology or medicine who were connected with administration or research in the KWG or the KWIs for basic science. The eligible population consists of all prizewinners associated with the central administration of the KWG or the KWIs until World War II; the award may have been made after that time.

The functions that the prizewinners performed in the society are indicated in the table. These ranged from being a member of a research staff to serving as president of the society. Max Planck probably wore more hats than anybody else during his half-century of service; by the time of World War II, he had been part of the society's Senate, served as president (1930–1937), and been on the boards of three KWIs. In 1945 to 1946, he briefly resumed the presidency until the society could be reconstituted. It was appropriately renamed the Max-Planck Society for the Advancement of Sciences (Max-Planck Gesellschaft zur Förderung der Wissenschaften).[9] Planck was typical not in the number of functions he filled, but in that these were generally administrative or advisory ones. Fewer than half of the 20 winners did scientific work in a KWI. Even this small number errs in excess because it includes Albert

---

[8] Ibid., pp. 33–53 and Bernhard vom Brocke, "Die Kaiser-Wilhelm-Gesellschaft in der Weimarer Republik: Ausbau zu einer gesamtdeutschen Forschungsorganisation (1918–1933)," in Vierhaus and vom Brocke, eds., *Forschung im Spannungsfeld von Politik und Gesellschaft*, p. 273.

[9] J. L. Heilbron, *The dilemmas of an upright man: Max Planck as spokesman for German science* (Berkeley, 1986), pp. 163–174 and 198–199.

Table 11. *Nobel prizewinners affiliated with the KWG and the KWIs in basic science: positions and prizework*

	Position	Dates of Prizework	Prize
Winner and position	(1)	(2)	(3)
KWG Central Administration			
Carl Bosch, senator	1922–1937	1909–[b]	Ch 1931
president	1937–[a]		
Paul Ehrlich, senator	1911–1915	1889–[b]	M 1908
Albert Einstein, senator	1922–1924	1905	Ph 1921
Emil Fischer, senator	1911–1919	1881–1891	Ch 1902
Fritz Haber, senator	1922–1933	1907	Ch 1918
Otto Hahn, senator	1928–1935	1938–1939	Ch 1944
Max von Laue, senator	1925–1929	1912	Ph 1914
Walther Nernst, senator	1919–1933	1906	Ch 1920
Max Planck, senator	1916–1930	1900–1901	Ph 1918
president	1930–1937		
J. H. van't Hoff, senator	1911	1884	Ch 1901
KWI for Biology			
Emil Fischer, *Kuratorium*[c]	1914–1919	1881–1891	Ch 1902
Otto Meyerhof, researcher	1924–1929	1919	M 1922
Hans Spemann, director[d]	1914–1919		M 1935
external researcher	1924–1935		
Otto Warburg, director	1913–1931	1924	M 1931
KWI for Chemistry			
Carl Bosch, *Kuratorium*	1921–[a]	1909–[b]	Ch 1931
Paul Ehrlich, *Kuratorium*	1911–1915	1889–[b]	M 1908
Emil Fischer, *Kuratorium*	1911–1919	1881–1891	Ch 1902
Otto Hahn, reseacher	1912–[a]	1938–1939	Ch 1944
director	1924–[a]		
Max von Laue, *Kuratorium*	1921–[a]	1912	Ph 1914
Walther Nernst, *Kuratorium*	1921–[a]	1906	Ch 1920
Max Planck, *Kuratorium*	1931–[a]	1900–1901	Ph 1918
Otto Wallach, *Kuratorium*	1917–1928	1890–[b]	Ch1910
Otto Warburg, *Kuratorium*	1921–[a]	1924	M 1931
Heinrich Wieland, *Kuratorium*	1921–[a]	1912–[b]	Ch 1927
Richard Willstätter, director	1912–1916	1905–1914	Ch 1915
Kuratorium	1921–1938		
external rcsearcher	1927–1933		
A. O. R. Windaus, *Kuratorium*	1926–1932	1925–1927	Ch 1928

Critical and empirical studies

Table 11 (cont.)

Winner and position	Dates of		
	Position (1)	Prizework (2)	Prize (3)
KWI for Experimental Therapy (Biochemistry)			
Adolph Butenandt, director	1936–[a]	1932	Ch 1939
Paul Ehrlich, *Kuratorium*	1912–1915	1889–[b]	M 1908
Hans von Euler-Chelpin, external researcher	1925–[a]	1905–[b]	Ch 1929
KWI for Physical Chemistry			
James Franck, researcher	1919–1920	1913–1914	Ph 1925
external researcher	1926–1938		
Fritz Haber, director	1911–1933	1907–[b]	Ch 1918
Richard Willstätter, *Kuratorium*	1923	1905–1914	Ch 1915
KWI Cell Physiology			
Otto Warburg, director	1931–[a]	1924	M 1931
KWI for Aerodynamics			
Max Planck, *Kuratorium*	1926–[a]	1900–1901	Ph 1918
KWI for Physics			
Albert Einstein, director	1917–1933	1905	Ph 1921
Peter Debye, director	1937–1939	1912–1929	Ch 1936
James Franck, *Kuratorium*	1931–1933	1913–1914	Ph 1925
Fritz Haber, director	1917–1933	1907–[b]	Ch 1918
Kuratorium	1917–1928		
Max von Laue, director	1922–[a]	1912	Ph 1914
Kuratorium	1922–[a]		
Walther Nernst, *Kuratorium*	1917–1931	1906	Ch 1920
Max Planck, *Kuratorium*	1917–[a]	1900–1901	Ph 1918
Erwin Schrödinger, *Kuratorium*	1931–1933	1925–1926	Ph 1933

[a] The absence of an end date means that the individual continued service in the KWG past 1939.

[b] The absence of an end date means that the prizework was ongoing. The dates of the prizework are approximate.

[c] *"Kuratorium"* is used as a blanket term to encompass all of the different steering groups of the KWIs (*Verwaltungsausschuss, Verwaltungsrat, Wissenschaftlichen Beirat*, and the like).

[d] *"Director"* is used to refer also to assistant director and head of department.

Einstein, as a director of the KWI for Physics, which did not have its own building or facilities until after he had left Germany. It also includes foreigners, such as the Swede Hans von Euler-Chelpin, who were long-term associate members of the institutes but excludes short-time visitors such as Eugene Wigner (Ph 1963), who spent 1932 at the KWI for Physical Chemistry.

Among the prizewinners engaged in research rather than administration, there were only a few – Otto Hahn, Hans Spemann, and Otto Warburg – who had done any of the work that brought them the prize while affiliated with a KWI. This was a consequence of the institutes being organized around scientists who had had distinguished careers in research. One subdirector (Max von Laue) came into the society as a laureate. Other directors had already done their prizewinning work and were close to nobelization when they took up their appointments. The prime example is Einstein, who became director of the KWI for Physics in 1917 and was awarded the Nobel prize in physics for 1921 for his "discovery of the law of the photoelectric effect." The calls of Fritz Haber and Richard Willstätter in 1912 to directorships of the KWI for Physical Chemistry and the KWI for Chemistry, respectively, coincided with their emerging as serious candidates for Nobel prizes in chemistry, which Haber won in 1919 (prize for 1918) and Willstätter in 1915.

It is not unlikely that the future nobelization of Haber played a role in his recruitment as director of the KWI for Physical Chemistry. In his memoirs, Friedrich Schmidt-Ott relates that he traveled to Sweden in the summer of 1910 to seek Arrhenius's advice on whom to appoint as director of the institute, which was founded the following year. Arrhenius strongly advised Schmidt-Ott to offer Haber the directorship.[10] They almost certainly also discussed Haber's chances for the Nobel prize inasmuch as the Nobel Committee for Chemistry was then deliberating which of several competing methods for the industrial production of nitrogenous fertilizers (the Birkeland-Eyde, the Schönherr, or the Frank-Caro process) would be proposed for an award. Although the Haber-Bosch process was still in the development stage, the insistence of some committee members that an award be made in this area clearly augured well for Haber.[11]

[10] Fritz Haber to Svante Arrhenius, September 23, 1911, Arrhenius Collection, KVA; Friedrich Schmidt-Ott, *Erlebtes und Erstrebtes, 1860–1950* (Wiesbaden, 1952).
[11] Crawford, *The beginnings of the Nobel institution*, pp. 181–183.

Critical and empirical studies

Haber's appointment benefited Arrhenius, whose rift with Walther Nernst had deprived him of a listening post at the Chemical Institute at the University of Berlin. The role of the intelligence gatherer, who provided information useful in the selection of Nobel prizewinners, fell to Haber and his institute, where Arrhenius became a frequent visitor.

The *Ruf* of nobelizable scientists to head up the KWIs reflected the society's policy to build each institute around its director. The institute served the director's scientific interests and provided all the assistance and equipment he desired; but it did not provide an environment in which subordinates had much chance of following their own line of research to nobelizable discoveries. Otto Warburg, for instance, made sure that he would be able to dominate all phases of the research process by restricting space at the work benches of his institute (KWI for Cell Physiology). He accepted very few doctoral students, relying for the most part on technicians trained in industry. He felt that, in contrast to academics, technicians knew their place and were not concerned with their careers, nor did they bother him with letters of recommendation.[12] Although Warburg may have been an extreme example, during the entire period studied here, only directors or assistant directors of the institutes were being nominated for the science prizes.

Another consequence of the same policy was a rapid turnover among the assistants and junior collaborators. At the KWI for Chemistry, for example, 17 Ph.D. candidates and postdoctorates entered, and most of them also left, Willstätter's department during his tenure (1912–1916); 25 did the same in Ernst Beckmann's during his 12 years (1911–1923); and 86 frequented the combined departments of Otto Hahn and Lise Meitner (1912–1939). Although the system did not produce any Nobel candidates, apart from the most senior staff, it did turn out a quantity of high-level work along the research lines opened by the directors.

That the KWG should seek past or potential prizewinners as directors of its institutes needs no special explanation. Nobelists, whether actual or potential, had the international certification of excellence that enhances local value: They attracted attention and also men and money to the society. Did they in turn exercise any particular influence on the general policy of the society? Certainly they did as individuals, but as laureates, as representatives of the value of basic research as increasingly

[12] Hans Krebs, *Otto Warburg*. Grosse Naturforscher No. 41 (Stuttgart, 1979), pp. 24–25.

emphasized by the Nobel institution, they were not notably successful. The best case in point is the Kaiser Wilhelm Institute for Physics that began to exist, but only on paper, in 1914. The majority of the physicists and physical chemists on the board of the KWI for Physics had either already won the prize or were about to do so (see Table 11). Einstein, the first director, was joined by von Laue as subdirector in 1922. Yet they did not succeed in putting up a building for the institute, thus giving it a real existence, until 1937. Their failure (for they did try) may appear all the more in need of explanation because the original plans for the KWG gave a physics institute pride of place.

All the physicists associated with the KWI for Physics from its beginning identified with basic research and even with theory. The emphasis on basic over applied work agreed with the Nobel institution but not with the KWG. Among the society's most powerful early supporters were chemists promoting an analogue for their purposes of the Imperial Institute of Physics and Technology (Physikalisch-Technische Reichsanstalt), which supported some basic research among a much larger quantity of applied work, particularly standardization of measuring units.[13] For decades chemistry had been more closely associated with industry than physics was, and consequently the Nobel Committee for Chemistry, and the university professors it consulted, were more familiar with and respectful of applied science than the Nobel Committee for Physics was. Several of the laureates connected with the KWI for Chemistry, particularly Carl Bosch, Emil Fischer, and Walther Nernst, had close ties with practical applications of chemistry.

During World War I, the KWIs for Chemistry and Physical Chemistry turned almost entirely to work related to the war effort, and most immoderately so in the case of Haber's institute. It became the army's center for gas warfare with a personnel of about 1,500 (of whom 150 were scientists) and was soon put under military command.[14] The emphasis on research of military and industrial interest, and the wide demonstration that science could have immediate payoffs, led to the creation of institutes devoted to metallurgy, textiles, and leather during

[13] Jeffrey A. Johnson, "Vom Plan einer Chemischen Reichsanstalt zum ersten Kaiser-Wilhelm-Institut: Emil Fischer," in Vierhaus and vom Brocke, eds., *Forschung im Spannungsfeld von Politik und Gesellschaft*, pp. 486–515.
[14] Lothar Burchardt, "Die Kaiser-Wilhelm-Gesellschaft im Ersten Weltkrieg (1914–1918)," in Vierhaus and vom Brocke, eds., *Forschung im Spannungsfeld von Politik und Gesellschaft*, pp. 164–166.

the war or in the immediate postwar era.[15] Hardly any of these new institutes of applied science contained nobelizable personnel. The Great War accelerated the tendency toward applied research and made it a mark of the KWG as a whole.

In the late spring of 1914, the sponsors of the KWI for Physics – Haber, Nernst, Planck, Heinrich Rubens, and Emil Warburg – got commitments from the society and the Koppel Foundation sufficient for a small building, equipment, and operating expenses if the state would make a modest annual subvention. It is instructive that the proposers referred to the service physics could render chemistry in support of their plan, and that the existence of the Physikalisch-Technische Reichsanstalt made their task difficult. They therefore expected to concentrate their effort on a few special themes in three-year research programs to be carried out, in large measure, at universities.

The Prussian minister of education supported their modest proposal. His colleague, the minister of finance, rejected it, because he saw nothing in the public interest that would justify its expense. The war then came and the planned institute was a victim of the society's *Gründungsstopp*, that is, its decision on August 12, 1914, not to create any new institutes for the duration of the hostilities. This decision was rescinded in 1917, when the KWG set aside 50,000 marks ($12,500) for operating expenses for the Physics Institute, but not for a building. Thus, a paper KWI for Physics came into existence, under Einstein's directorship, for the novel purpose – for the KWG – of making small grants to university researchers.[16]

The KWG never did put up the money for a physics institute. Although it greatly increased the scope of its operations during the 1920s, the society found nothing for a building for physics that might serve the purposes of Einstein and his co-administrators and fellow prizewinners, von Laue and Planck. At last, the Rockefeller Foundation agreed in 1930 to provide the money ($655,000) for a small building for the KWI for Cell Physiology and a much larger one for the KWI for Physics. Cell Physiology was set up specifically for Otto Warburg, whose research on the metabolism of cancer cells showed promise for medical applications. The institute building was completed in 1931, the year Warburg

---

[15] Ibid., pp. 183–191; and vom Brocke, "Die Kaiser-Wilhelm-Gesellschaft in der Weimarer Republik," pp. 241–249.
[16] Burchardt, "Die Kaiser-Wilhelm-Gesellschaft im Ersten Weltkrieg (1914–1918)," p. 177.

won the Nobel prize in medicine for his earlier work on the respiratory enzyme.[17]

The KWI for Physics, which had the heritage of the past paper institute to carry and still showed no practical utility, foundered. Planck, now president of the KWG, decided to give the new institute an experimental cast, and to try to lure a Nobel laureate, James Franck (Ph 1925) from Göttingen to direct it. Franck was promised to succeed Nernst, who was about to retire from the chair of experimental physics at the University of Berlin. When all the necessary arrangements had been made, however, the Nazis came to power and Franck left Germany.

The political situation made the Rockefeller Foundation seriously consider withdrawing its offer to finance the construction of the institute. It was largely through a personal appeal, based as much on his reputation for probity as on his eminence in science, that Planck obtained the release of the foundation's funds. Having secured another Nobel laureate, the Dutchman Peter Debye (Ch 1936), as director, convinced the Nazi government to pay for the operating costs. The institute, now named the Max Planck Institute for Physics, was inaugurated in 1938. When the war broke out, the institute was taken over by the military for work on nuclear fission, and Debye as a non-German had to leave.[18]

This saga seems to show that the prestige of prizewinners did not bring with it proportional influence in the KWG's affairs. The swerve toward applied research in World War I was followed by an increasing dominance of the society's affairs by industry. This orientation was confirmed when Nobel laureate Carl Bosch (Ch 1931), who was closely associated with I. G. Farben, succeeded Planck as president in 1937. In this environment an institute for pure physics, despite the impressive array of prizewinners backing it, could not make good its claims for adequate support from the society.

Nominees

Table 12 lists everyone who, when affiliated with the KWG or a KWI, received a nomination for a Nobel prize in physics or chemistry. The table gives the total number of nominations received by each candidate

[17] Krebs, *Otto Warburg*, pp. 25–26.
[18] Heilbron, *Dilemmas of an upright man*, pp. 94–96, 175–179; Kristie Macrakis, "The Rockefeller Foundation and German physics under national socialism," *Minerva* 27 (1989): 33–57.

Table 12. *KWG and KWI candidates in physics and chemistry and the nominations they received during their affiliation*

Candidate and discipline	Dates of nomination (1)	Total nominations (2)	Nominations received from	
			Germans (3)	KWG personnel[a] (4)
Physics				
Albert Einstein	1917–1922	52	26	10
James Franck	1923–1926	14	8	4
Friedrich Paschen	1923–1933	32	31	17
Max Planck	1916–1919	20	18	5
Ludwig Prandtl	1928–1937	4	3	0
Erwin Schrödinger	1931–1933	21	6	3
Emil Warburg	1929	1	1	1
Chemistry				
Max Bodenstein	1929–1937	15	14	0
Karl Friedrich Bonhoeffer	1930	1	0	0
Carl Bosch	1926–1931	14	5	1
Adolph Butenandt	1936–1939	4	0	0
Nikodem Caro	1932–1933	2	2	2
Theodor Curtis	1913–1922	15	15	1
Paul Ehrlich	1911	1	1	0
Hans von Euler-Chelpin	1924–1929	16	14	10
Emil Fischer[b]	1916–1919	2	2	1
Herbert Freundlich	1921–1928	2	0	0
Fritz Haber	1912–1919	10	6	2
Carl Harries	1913–1917	7	7	2
Carl Neuberg	1929–1935	17	17	5
Wilhelm Schlenk	1924–1929	3	1	0
Alfred Stock	1929	1	0	0
Otto Warburg	1929–1931	2	1	0
Heinrich Wieland	1924–1934	12	12	1
Richard Willstätter	1912–1915	24	14	8
A. O. R. Windaus	1926–1928	7	6	2
Physics and chemistry				
Otto Hahn	1914–1939	19	13	8
Lise Meitner	1924–1939	14	11	8
Walther Nernst	1917–1921	37	22	2

[a]"KWG personnel" are nominators who at any time were affiliated with the KWG.
[b]Although already a laureate (Ch 1902), Emil Fischer was nominated for a second prize in chemistry in 1916 and 1919.

and shows, first, how many were issued by German nominators, and second, how many of these came from people associated with the KWG or one of the KWIs included in this study. Before discussing these matters, one general point should be made about the table: Nominees in chemistry and physical chemistry (21 in all, including Paul Ehrlich and Emil Fischer, both nominated for a second prize) greatly outnumbered those in physics (7).

The surplus of nominations in chemistry, and even in chemistry close to practice was a consequence of the society's own priorities in basic science. Together the senior staffs of the KWI for Chemistry and the KWI for Physical Chemistry numbered 12 in 1920 to 1922 and 15 in 1928; and their swarms of assistants (not including *Doktoranden*) numbered, for the same years, between 25 and 30. The KWI for Physics had no employees but its director and subdirector and no committees affiliated with it, and its board was composed of senior (and mostly nobelized) scientists. Furthermore, the farther one moved from basic science, the fewer the nominations. Very few, if any, nominees or nominators, let alone a prizewinner, were on the staffs of KWIs in applied science, for example, the institutes for research in metallurgy, leather, and textiles. These institutes were oriented still more toward practice than the KWI for Aerodynamic Research, which garnered four nominations for its director, Ludwig Prandtl during the period between 1928 and 1937. That few if any of the personnel of the KWIs for applied science figure on the Nobel lists is an indication both of the policies with respect to Nobel selections and of the impertinence of the claims of KWG spokesmen, who, when it suited their purpose, insisted that all the institutes pursued basic research.

The predominance of chemists also explains the meager number of nominations that the KWG-KWI nominees received from outside Germany. As discussed in the study of nominating patterns in the wake of World War I (see Chapter 3), international exchanges in chemistry suffered more from the war, and for a longer period than those in physics. The chemists associated with the KWIs received even fewer nominations, however, than their counterparts at other German institutions. This was no doubt due to the intersection of the war with the policies of the KWG.

Because the society's nobelizable personnel between 1911 and 1939 were almost all established scientists before the war, they had an opportunity to compromise themselves in the eyes of their former friends and colleagues in the entente countries by chauvinistic declarations. The war work of the society and its scientists, particularly the making of poison gas in the KWI for Physical Chemistry under the directorship of Haber, doubtless made it even less likely that Allied chemists would support their counterparts in the KWG. Again, owing to the continued interruption of normal international scientific relations after the war, the younger workers in the institutes, who in any case would have had a hard time gaining attention, had fewer opportunities for international contacts and recognition. More important, the new generation of physicists, who invented quantum mechanics in the 1920s and who attracted a fair amount of nominations from outside Germany, had hardly any ties with the KWG.

The KWG and KWI candidates who received nominations from people affiliated with the society are listed in Table 12 (column 4). Such support was more conspicuous for members of the senior research staffs – von Euler-Chelpin, Hahn, Meitner, Willstätter – than for those who held only decision-making or advisory posts in the society (Planck, Schrödinger, Nernst). Haber is the most prominent example of a prizewinner, and a senior member of the research staff to boot, who received hardly any support from society nominators. In 1918, the year for which Haber received the prize, his fellow chemist in the society, Wilhelm Schlenk, gave his vote to Frederick Soddy! The indifference to Haber's claims by scientists in the society he graced may be connected with his work on poison gas, with which the KWG nominators might not have been eager to identify.

Nominators

Table 13 lists those who, when affiliated with the central administration of the KWG or a KWI, nominated anyone for a Nobel prize in physics or chemistry. By statute, all previous prizewinners have a permanent right to nominate, and most of those from the KWG exercised their franchise at least once, some of them many times. The nonprizewinners came under the statutory provision that gave the Swedish Academy of Sciences a mandate to invite individuals and institutions to suggest candidates for a specific year. KWG nominators who proposed candidates under this provision were invited not as affiliates of the society

Table 13. *KWG and KWI nominators in physics and chemistry and their candidates, 1911–1937[a] during their affiliation with the KWG*

Nominator	Dates of nomination (1)	Total nominations (Nominations made to KWG personnel) (2)	Nominations made to prizewinners (3)
Ernst Beckmann	1914–1922	7 (7)	4
Max Bodenstein	1929	1 (1)	1
Paul Ehrlich	1912–1914	5 (5)	4
Albert Einstein	1919–1933	7 (3)	7
Hans von Euler-Chelpin	1923–1937	14 (1)	5
Emil Fischer	1911–1919	9 (4)	6
James Franck	1927–1933	34 (5)	20
Herbert Freundlich	1929	1 (0)	1
Fritz Haber	1920–1933	5 (2)	2
Otto Hahn	1929–1937	2 (0)	1
Arthur Rudolf Hantsch	1920	1 (1)	1
Ludwig Knorr	1912	1	1
Max von Laue	1921–1937	22 (9)	14
Max Le Blanc	1927	4	1
Kurt Meyer	1921	1 (1)	1
Walther Nernst	1923–1933	19 (5)	5
Carl Neuberg	1929	2 (1)	1
Friedrich Paschen	1926	2 (0)	1
Max Planck	1917–1937	34 (19)	17
Ludwig Prandtl	1913	1 (1)	1
Wilhelm Schlenk	1918	1 (0)	1
Franz Eilhard Schulze	1915	1 (1)	1
Alfred Stock	1922	1 (0)	0
Hermann Thoms	1915	1 (1)	1
Otto Wallach	1912–1914	1 (0)	0
Karl Hermann Wichelhaus	1922	1 (1)	0
Heinrich Wieland	1926–1937	15 (4)	8
Richard Willstätter	1916–1936	16 (11)	8
A. O. R. Windaus	1929–1932	2 (0)	2

[a]After the peace prize was awarded to Carl Ossietzki in 1936, Hitler ordered all Germans to boycott the Nobel prize. This order was followed to the letter. No scientist in Germany invited to nominate candidates for the prizes did so in 1938 or 1939.

but as chairholders or honorary professors in physics and chemistry at a university or *technische Hochschule*. Relatively few institute directors and even fewer workers in the institutes below the directors received an invitation. For example, Herbert Freundlich, Hahn, and Carl Neuberg were consulted, or to be more accurate, exercised a franchise, only once or twice. It is curious that these infrequent nominators tended to look beyond Germany for candidates: Freundlich's single vote went to an American, two of Hahn's three votes to a Dutchman and a Hungarian, respectively, and Neuberg's to a Swede and an Englishman.

Not surprisingly, the KWG nominators who were already prizewinners and therefore had permanent nominating rights accounted for the majority of nominations. Here Planck and Franck head the list with 34 nominations each, followed by von Laue and Nernst with, respectively, 22 and 19. All these nominators chose primarily Germans. In this they did not differ from their fellow laureates outside the society or, indeed, from German nominators in general. In the period 1916 to 1937, only about 10 percent of the overall German vote went to non-German candidates. Nobel laureates were no different from nonwinning candidates in choosing primarily their own compatriots. (See Chapter 3.)

Most of the active nominators (those who made from five to thirteen choices) gave a generous fraction of their suffrage to KWG people. Thus, Beckmann, seven of seven, Ehrlich, five of five, Einstein, three of seven, Fischer four of nine, and Willstätter eleven of sixteen. Among the most active nominators (twenty or more) the fraction was less: five of thirty-four with Franck, nine of twenty-two with von Laue, and nineteen of thirty-four with Planck. The most active nominators were also the most successful in nominating laureates if one considers that for each additional nomination the probability of picking a prizewinner would seem to decline. Here, it is worth noting, the unsurpassed performance of Planck, half of whose 34 nominations made over a quarter-century concerned prizewinners. This coincidence again points up that the KWG did not provide much opportunity for, or strive to bring forward, members who were not patently nobelizable.

Conclusions

The type of elite science promoted by the KWG clearly made much room for Nobel prizewinners. The roles they performed in and for the society give a glimpse of some of the particulars of the elite conception of scientific organization that characterized the KWG. The roles they performed in the Society can be summarized as follows:

1. Laureates were more likely to adorn the administrative and advisory organs of the society than to have carried out their prizewinning work in one of its institutes. This was a result of the general emphasis on power and prestige as a means of attracting resources to the organization. Furthermore, in addition to giving the society an international standing, the bevy of laureates conferred a sense of national mission on the KWG, raising it above the universities, which were basically local institutions.

2. The scientists – Einstein, Haber, von Laue, and Willstätter – who received the *Ruf* to head up one of the KWIs had either already received the prize or were about to do so. This reflected the society's policy to build each institute around a prominent scientist, who would produce important results, it was thought, if given the time and resources to devote himself entirely to research. While generally carried out on a high level – as evidenced, for example, by the achievements of the Hahn-Meitner team – the continuity of research in the KWIs may have suffered from the gap that separated the nobelized and nobelizable personnel from the rest. This proposition is difficult to test, however, because of the disruptions caused in the research programs of the KWIs by world historical events.

3. There was "inbreeding" of nominations in the sense of KWG nominees receiving more support both from their fellow Germans and their KWG colleagues than did other German nominees. Such corporatist attitudes obviously ran counter to the objectives of the Nobel institution, which stood for universalism and internationalism in science. It was in keeping, however, with the nationalistic spirit that had inspired the creation and growth of the society.

Still, for all the prestige that the prizes conferred on KWI directors, this did not translate into proportional influence over decision making in the society. Their roles in the society qua prizewinners did not go much beyond that of being symbols of international approbation and distinction. This is demonstrated by the failure of the galaxy of prizewinners associated with the KWI for Physics to raise sufficient support within the society to make their paper institute operational. The resistance that Planck, von Laue, and Einstein encountered in trying to set up their institute may be traced to the increasing emphasis on applied science by the society's administration and industrial supporters in the interwar period.

In important respects, the role of the prizewinners for the KWG, when not simply that of enhancing KWG power and prestige, related to the tension between basic and applied research. Here, the prize-winners, whose award citations almost without exception emphasized basic science, could attract the attention, and perhaps also the support, of elements opposed to the society's drift toward application. They constituted a living and sometimes effective reminder of the original purpose of the KWG. An example is the consecration of Haber in 1919 (prize for 1918), despite his perversion of his science to gas warfare. Haber's leadership of the KWI for Physical Chemistry, whose first objective before the war had been basic research, probably helped to mitigate resistance to his selection. Reciprocally, his prize hastened the return of his institute to its basic research mission.

Again, the presence of the prizewinners may have helped to soothe industrialists impatient at the pace of applied work in the KWI for basic research they supported. An example would be the KWI for Chemistry, where the status of Hahn, Meitner, and other nobelizables may have provided some compensation to industrialist contributors for their unfulfilled hopes for results useful to practice. Overall, these examples show how the symbolic, reputational uses the prizes served in the KWG were related above all to alleviating the tension between the basic and applied research that characterized the society after World War I.

6

~~~~~~

Nobel laureates as an elite in American science

It is tempting to view scientific development in terms of the inherent elitism of the enterprise, and, as we have seen in Chapter 1, studies of scientific elites have constituted a nonnegligible strand of work in the social history of science. This for several reasons: First of all, the idea that scientists constitute a chosen few, who by virtue of superior ability and dedication to the task stand high in society, agrees more or less with the facts. Before World War II, there were indeed few practitioners of science and they were generally accorded high social status, although not always as high as at present.[1] Second, scientists themselves believe that a certain amount of inegalitarianism is necessary to advance the enterprise, and some have not hesitated to express this. "There is no democracy in physics" is the oft-quoted statement of Luis Alvarez, Nobel laureate (Ph 1968). "We can't say that some second-rate guy has as much right to an opinion as Fermi."[2] Einstein elevated this communion of superior minds into a principle of universalism in science. "The supranational character of scientific concepts and language," he wrote, "is due to the fact that they were formed by the greatest brains of all countries and all times."[3] Third, elitism has been rendered visible, and perpetuated, through the elaborate system honoring scientific achievement that has evolved over time. The plethora of prizes and other honors used to recognize and reward scientific achievement are the insignia of

---

[1] In 1947, "scientist" and "nuclear physicist" were ranked eighth and eighteenth, respectively, in terms of prestige of some 90 occupations, by a sample of the American adult population. In 1963, they shared third place in a similar survey. Robert W. Hodge, Paul M. Siegel, and Peter H. Rossi, "Occupational prestige in the United States: 1925–1963," *American Journal of Sociology* 70 (1964): 286–302.
[2] Quoted in Daniel S. Greenberg, *The politics of pure science* (New York, 1968), p. 43.
[3] Albert Einstein, *Conceptions scientifiques, morales et sociales* (Paris, 1952), p. 149.

scientists' standing in the eyes of their peers. The distribution of such honors within the scientific community has followed the principle referred to by Robert K. Merton as the "Matthew effect" in science, for it seems that those who already have been awarded such honors are also the ones most likely to receive new ones.[4]

Not surprisingly, the notion of a scientific elite marked off by accumulated advantages has prospered in the United States with its decentralized, mixed public and private university system and its corps of professional researchers competing for positions and resources in the academic marketplace. It is to the United States, then, that one has to turn for a close examination of the role that elitism has played in scientific development. Doing so naturally leads to the questions of how elitism has squared with the egalitarian and pluralistic assumptions of American democracy – these questions are clearly of too tall an order to confront, let alone answer, in a short study.[5] The much more limited objective of our inquiry here is to give empirical substance to the notion of an American scientific elite in the crucial period from the turn of the century to World War II, a period during which the American scientific enterprise moved to occupy front rank internationally. Was there a scientific elite? And if so, who were its members? What are the criteria that permit us to mark off some scientists and to collectively designate them as a national elite? What functions did the elite perform when it came to organizing the national scientific enterprise?

The American Nobel population as a national scientific elite

The most systematic attempt to give substance to the notion of an American scientific elite is found in a series of studies conducted during the 1960s and 1970s by the director of the Columbia Program in the Sociology of Science (Robert Merton) and participants in the program (in particular, Jonathan and Stephen Cole and Harriet Zuckerman). In

---

[4] Robert K. Merton, *The sociology of science: Theoretical and empirical investigations* (Chicago, 1973), p. 446.
[5] Cf. Daniel J. Kevles, *The physicists: The history of a scientific community in modern America* (New York, 1978); Roger L. Geiger, *To advance knowledge: The growth of American research universities, 1900–1940* (Oxford, 1986); and Ronald C. Tobey, *The American ideology of national science, 1919–1930* (Pittsburgh, 1971).

general, these studies viewed stratification and accumulation of advantages as immanent features of the American scientific enterprise in the twentieth century. Moreover, these traits were seen as functional to progress in science.[6] Scientists' rankings of prizes and other honors (election to academies of science, honorary doctorates, and the like) constituted the objective measures of social stratification in science. When the Coles asked about 1,300 physicists at American universities to rank 98 honorific awards according to their prestige, the Nobel prizes received the highest score.[7] The prizes figured most prominently in Zuckerman's work, which made the ultra-elite in American science coterminous with its Nobel laureates in physics, chemistry, and physiology or medicine between 1907 and 1972. It is probably not coincidental that this celebration of the Nobel prizes appeared in 1977, the year after America had swept *all* the prizes, including that for literature.[8]

Zuckerman argued that the American scientific enterprise in the twentieth century had a pyramidal structure as a result of stratification and accumulated advantage. To demonstrate this, she chose to study Nobel laureates in the United States, whom she saw as occupying the very top of the pyramid. Their small number (72) clearly made them more accessible for intensive study than the some 500,000 self-defined scientists who figured in the U.S. Census of 1974, although to sustain her argument, strictly speaking, she would have had to show that the mass of scientists at the base of the pyramid were indeed different from those at the top. The attributes she examined for the small group at the top were those relating to training, careers, scientific work, and recognition through honors other than the Nobel prize. On all these counts, she found that the laureates were set apart from the some 1,000 scientists elected to the U.S. National Academy of Sciences up to 1975, this being the group she had identified as occupying the next level of the pyramid. She could conclude, then, that "the laureates' careers . . . square rather well with the model suggested by the accumulation of advantage: the

[6] Merton, *Sociology of science*; Jonathan R. Cole and Stephen Cole, *Social stratification in science* (Chicago, 1973); Harriet Zuckerman, *Scientific elite: Nobel laureates in the United States* (New York, 1977).
[7] Cole and Cole, *Social stratification*, pp. 270–275.
[8] The 1976 prizewinners were Burton Richter and Samuel C. C. Ting (physics); William N. Lipscomb, Jr. (chemistry); Baruch S. Blumberg and Carleton D. Gajdusek (physiology or medicine); Saul Bellow (literature); and Milton Friedman (economics). The peace prize was reserved.

spiraling of augmented achievements and rewards for individuals and a system of stratification that is sharply graded."[9]

One way to find out if the laureates did indeed represent a breed apart is to compare them with the nonwinning candidates, Zuckerman having been unable to do so because of the secrecy rules still in force at the time of her study.[10] This comparison will be carried out using the same attributes – training, careers, scientific work, and recognition through honors other than the Nobel prize – that Zuckerman drew on to define the ultra-elite. It will concern 40 candidates for the prizes in physics and chemistry, 1901 to 1939, 13 of whom were prizewinners (see Table 14).

Before giving the results of this comparison, I will present a descriptive survey of how American physicists and chemists related to the Nobel institution in their capacity as nominees and nominators during the period up to World War II. This background is essential because Zuckerman treats her population as an ahistoric entity, viewing it in a socioinstitutional context restricted to the time she carried out her study.

American involvement with the Nobel prizes in physics and chemistry, 1901–1939

The beginnings of the American involvement with the Nobel institution were inauspicious. The only nominations for Americans in 1901, both in physics, had been handed in by nominators who proposed themselves, and were hence disallowed.[11] The one was by and for Henry Rowland, who perhaps was putting into practice for himself the best-science elitism he had advocated for America. The other was by and for R. H. Thurston, head of the Sibley College of Mechanical Engineering at Cornell University.

In the following two years there were no American candidates for the

[9] Zuckerman, *Scientific elite* p. 249.
[10] Zuckerman used an approximate list of candidates, mainly biomedical scientists, drawn up on the basis of the indiscretions that had been committed in the essays on the prizes in physiology or medicine and chemistry in the 1962 edition of the official history of the institution (Henrik Schück et al., eds., *Alfred Nobel: The man and his prizes* [Amsterdam, 1962]). There were too few Americans among these "occupants of the forty-first chair" (a reference to the 40 chairs of the French Academy) to permit their being systematically compared with the laureates with respect to elite attributes.
[11] Paragraph 7 of the statutes of the Nobel Foundation (1900) stated: "A direct application for a prize will not be taken into consideration."

Nobel laureates as an elite

Table 14. *American candidates for the Nobel prizes in physics and chemistry, 1901–1939*

	Dates of	
	Nomination	Prize
John Jacob Abel (1857–1938)	1925–1930	
Carl David Anderson (1905–1991)	1934–1936	Ph 1936
Carl Barus (1856–1935)	1920–1921	
Ira Sprague Bowen (1898–1973)	1930–1934	
Percy Williams Bridgman (1882–1961)	1919–1946	Ph 1946
William Wallace Campbell (1862–1938)	1901	
Arthur Holly Compton (1892–1962)	1925–1927	Ph 1927
William David Coolidge (1873–1975)	1922–1935	
Clinton Joseph Davisson (1881–1958)	1929–1937	Ph 1937
Edward Curtis Franklin (1862–1937)	1919–1921	
Lester Halbert Germer (1896–1971)	1929–1937	
William Francis Giauque (1895–1982)	1936–1937	Ch 1949
Woolcott Gibbs (1822–1908)	1902	
Moses Gomberg (1866–1947)	1915–1938	
George Ellery Hale (1868–1938)	1909–1934	
William Draper Harkins (1873–1951)	1933–1938	
William Francis Hillebrand (1853–1925)	1913–1914	
Arthur Edwin Kennelly (1861–1939)	1935	
Irving Langmuir (1881–1957)	1916–1932	Ch 1932
Ernest Orlando Lawrence (1901–1958)	1938–1939	Ph 1939
Gilbert Newton Lewis (1875–1946)	1922–1935	
Alfred Lee Loomis (1887–1975)	1937	
Theodore Lyman (1874–1954)	1918–1926	
Arthur Michael (1853–1942)	1917–1922	
Albert Abraham Michelson (1852–1931)	1904–1907	Ph 1907
Robert Andrews Millikan (1868–1953)	1916–1926ª	Ph 1923
Edward Williams Morley (1838–1923)	1902–1910	
Harmon Northrop Morse (1848–1920)	1903–1910	
Arthur Amos Noyes (1866–1936)	1920–1927	
Fredrick Belding Power (1853–1927)	1923–1925	
Michael Idvorsky Pupin (1858–1935)	1929	
Theodore William Richards (1868–1928)	1902–1915	Ch 1914
Henry Augustus Rowland (1848–1901)	1901	
Fernando Sanford (1854–1948)	1918–1920	
Carl Frederick Schmidt (1893–?)	1930	
Wendell Meredith Stanley (1904–1971)	1938–1939	Ch 1946
Otto Stern (1888–1969)	1925–1939	Ph 1943

Table 14 *(cont.)*

| | Dates of | |
	Nomination	Prize
Harold Clayton Urey (1893–1981)	1934	Ch 1934
Edward Wight Washburn (1881–1934)	1934	
Robert Williams Wood (1868–1955)	1914–1937	

Note: Included here are the candidates who received two or more nominations. A few inventors and technologists (Edison and the Wright brothers) who received more than one vote are excluded. Three–Carl Barus, Henry Rowland, and Woolcott Gibbs, from Kevles's list of productive physicists and chemists (see this chapter, note 22)–are included, although they received only one nomination.
ªR. A. Millikan was nominated for a second prize in 1926.
Source: Elisabeth Crawford, J. L. Heilbron, and Rebecca Ullrich, *The Nobel population, 1901–1937: A census of the nominators and nominees for the prizes in physics and chemistry* (Berkeley and Uppsala, 1987).

physics prize. In chemistry, Gibbs – not Josiah Willard but Woolcott, the grand old man of American chemistry, then in his eighties – was put forth in 1902 by C. E. Munroe of the Columbian University of Washington, who called him "by far the most distinguished of American chemists." In addition, Munroe nominated two younger chemists, Edward W. Morley and Theodore W. Richards, for their determinations of atomic weights.

The physics prize awarded A. A. Michelson in 1907 for his interferometer was the first for an American citizen, but it was also an American prize in that the work that was rewarded, the construction and use of precision instruments, dominated physics research in the United States at the turn of the century. That Michelson received the prize with very few nominations was largely due to the strong position of measuring physics in Sweden, particularly in the Uppsala physics department, whose members made up the majority of the Nobel Committee for Physics.[12]

The period up to World War I has been called the "Michelson era"

[12] Elisabeth Crawford, *The beginnings of the Nobel institution: The science prizes 1901–1915* (Cambridge and Paris, 1984), pp. 56–59, 173–174; Robert Marc Friedman, "Americans as candidates for the Nobel prize: The Swedish perspective" in Stanley Goldberg and Roger H. Stuewer, eds., *The Michelson era in American science, 1870–1930* (New York, 1988), pp. 272–287.

in American physics; in Nobel selections, once Michelson had received the prize, it became the "Hale era." George Ellery Hale was nominated for the first time in 1909, received strong support for each of the five years 1913 through 1917, and was proposed intermittently until 1934. The great advances made in astrophysics through Hale's invention of the spectroheliograph put him in a strong position for a prize because they found favor with both the specialized precision physicists on the committee and those more broad-minded like Svante Arrhenius, who had an interest in merging micro- and macrophysics into one large entity, cosmic physics. The practicalities of Hale's award – Should he receive the prize alone? Together with Henri Deslandres, who had contributed significantly to the development of the spectroheliograph? Or perhaps with still other astrophysicists? – presented the committee with so many problems, however, that although favorably disposed, the matter was put in abeyance until after World War I. When taken up again in the early 1920s, committee priorities had changed, and astrophysics was pronounced no longer part of physics.[13]

The involvement of American physicists changed character in the interwar period as the quantity and quality of work and the institutional support the work received were put on a par with and eventually came to surpass those of major scientific powers, Germany in particular. Atomic and nuclear physics were the beneficiaries of much of this support; they were also the areas of most interest to the Nobel Committee for Physics. It is not surprising, then, that the string of awards made to Americans after World War I concerned those areas.

The first postwar prize went to Robert Millikan in 1923 for his oil-drop experiments establishing the elementary charge of the electron. It was followed by those to Arthur H. Compton (Ph 1927), Carl D. Anderson (Ph 1936), Clinton J. Davisson (Ph 1937), and Ernest O. Lawrence (Ph 1939). In contrast to the prewar years, alongside the prizewinner and the lone nonwinning candidate, there was now a field of contenders. Some of them, Lester H. Germer and Ira S. Bowen, for instance, were cited for collaborations with the eventual prizewinner, an indication of the new importance of teamwork in American science.

Another important change in the involvement of American physics with the Nobel institution was the international nominator support given

---

[13] Friedman, "Americans as candidates," pp. 278–282.

not just the winners but also such a perennial nonwinner as Robert W. Wood. Early candidates such as Michelson, Hale, and Millikan had garnered only one-quarter of their nominations from outside America. By contrast, for Davisson, Anderson, and Wood, outside support accounted for close to eight-tenths of their vote. Outside support was particularly strong for E. O. Lawrence, who won the physics prize of 1939 for "the invention and development of the cyclotron," as the official citation read. The campaign waged by the Compton brothers for Lawrence in 1939 brought forth nominations from former American prizewinners, but the support of leading international atomic and nuclear physicists (Niels Bohr, Enrico Fermi, E. Amaldi, and C. V. Raman, among others) was no doubt decisive for the final outcome. Lawrence's prize highlighted the importance of instrumentation in nuclear physics, but above all, it confirmed the privileged position that his laboratory had assumed nationally and internationally when it came to building and operating cyclotrons.[14]

The chemists' involvement with the Nobel institution reflects the more eclectic nature of work in this discipline and its broader institutional base. The names of the candidates also show that the discipline was more diverse than is indicated by the five prizewinners whose candidacies were put forth before World War II: T. W. Richards, Irving Langmuir, Harold C. Urey, William F. Giauque, and Wendell M. Stanley. In physical chemistry, for instance, the names of the award winners, Langmuir and Giauque, should be supplemented with those of William D. Harkins, who made important contributions to surface chemistry, and Gilbert N. Lewis, "the only chemist in America who ranked with Irving Langmuir."[15] Lewis garnered a record 35 nominations during the 13 years he was a candidate for the prize he never won.

In inorganic chemistry, T. W. Richards was preceded in his lifework of determining atomic weights by E. W. Morley. Another candidate who did important work on atomic weights was a government scientist, William F. Hillebrand, who served both in the U.S. Geological Survey (1880–1908) and in the U.S. Bureau of Standards, where he was chief chemist (1908–1925).

Although no American organic chemist won a Nobel prize before

---

[14] J. L. Heilbron and Robert Seidel, *Lawrence and his laboratory: A history of the Lawrence Berkeley Laboratory*, vol. 1 (Berkeley, 1990), pp. 484–493.
[15] Kevles, *The physicists*, p. 225.

World War II, this was not for a lack of candidates. Among them were Arthur Michael, "the best organic chemist in the United States," who worked not at one of the research universities that were emerging in the late nineteenth century but at Tufts College;[16] Edward C. Franklin, who built up the chemistry department at Stanford University; and Moses Gomberg, at Michigan, who specialized in the chemistry of free radicals. Finally, the population of candidates includes the biochemist Wendell M. Stanley, whose 1946 prize for his work isolating the tobacco mosaic virus in pure crystalline state inaugurated the string of awards to Americans for work in this speciality after World War II.[17]

The American Nobel population also included a handful of inventors and technologists, none of whom won prizes (the prize juries tended to favor basic research and academic scientists). The men who probably did the most to change American and world technology in the twentieth century – the Wright brothers and Thomas Alva Edison – were candidates for the prizes.[18] Other nonwinning candidates whose work had important practical applications were William D. Coolidge, who invented the high-vacuum, high-voltage, heated X-ray tube; Arthur E. Kennelly, who together with Oliver Heaviside postulated the reflecting atmospheric layer of ionized gases (later confirmed experimentally by E. V. Appleton), which proved to be of prime significance for radio transmission; and Michael I. Pupin, whose work on electrical resonators led to the finding that the insertion of inductance coils in telephone lines ("pupinized" lines) improved their performance.

All in all, the candidates represented, if not the totality of physicists and chemists who made significant contributions to the growth of their disciplines, at least a significant portion of them. Almost all the physics candidates, for instance, figure in Daniel Kevles's *The Physicists*, which draws the broadest picture of how the research profession came to maturity before World War II. It seems particularly interesting, then, to consider whether there were significant differences between the winners and the nonwinning candidates with respect to the attributes studied by Zuckerman.

[16] Daniel J. Kevles, "The physics, mathematics, and chemistry communities: A comparative analysis," in Alexandra Oleson and John Voss, *The organization of knowledge in modern America* (Baltimore, 1979), pp. 139–172.
[17] Lily E. Kay, "W. M. Stanley's crystallization of the tobacco mosaic virus, 1930–1940," *Isis* 77 (1986): 450–472.
[18] Crawford, *The beginnings of the Nobel institution*, pp. 142–143, 165–166.

Critical and empirical studies

Comparison of laureates and nonwinning candidates

In the following discussion, the laureates and the nonwinning candidates are compared with respect to the four attributes that Zuckerman used to delineate her ultra-elite of prizewinners: education and training, apprenticeship served under a laureate master, first jobs and subsequent employment, and rewards in the form of honors.

Education and training

Zuckerman's demonstration of her thesis that laureates are set apart by the advantages they accumulate throughout their careers naturally started with their education. She found that "the clumping of future members of the scientific ultra-elite in elite institutions begins early in the selective educational process." Whereas all the laureates she studied "went to college, of course,"[19] this was also true for eight-tenths of the nonwinning candidates. The ones without college education generally belonged to the generation of self-taught scientists born in the middle of the nineteenth century. More important, both Zuckerman's population of laureates for the three prizes (1907–1972) and my more restricted one of candidates (1901–1939) are alike in that more than half attended Ivy League or other elite colleges.[20]

As the future members of the elite moved from undergraduate to graduate education, the laureates and the nonwinning candidates converged on the 13 elite universities that granted degrees to 85 percent of Zuckerman's population of laureates, 83 percent of the nonwinning candidates, and 80 percent of the members of the National Academy of Sciences elected from 1900 through 1967.[21] Still, there are important differences, for these data refer only to those who, one, held doctorates and, two, had been granted these by American educational institutions. Among the laureates studied by Zuckerman, all with the exception of A. A. Michelson, held doctorates, and 80 percent had earned them in the United States. Almost half of the 18 laureates who held foreign

---

[19] Zuckerman, *Scientific elite*, p. 83.
[20] Ibid., pp. 82–86.
[21] The 13 elite universities on Zuckerman's list for the period 1920 to 1940 were California at Berkeley, California Institute of Technology, Chicago, Columbia, Cornell, Harvard, Illinois, Johns Hopkins, Massachusetts Institute of Technology, Michigan, Princeton, Wisconsin, and Yale.

doctorates were émigré scientists who had fled fascism in Europe. Of the candidates, both winners and nonwinners, half held doctorates from American institutions, one-fourth from foreign ones, and another one-fourth had no doctorate at all. The higher portion of those holding foreign doctorates or none at all is explained by the population of non-winning candidates going back to the late nineteenth century. At that time, a doctorate was not yet the sine qua non for an academic career that it had become early in the twentieth century, when the future laureates of the interwar period, the real starting point of Zuckerman's study, received their degrees.

This is shown most clearly when the candidates, winners and non-winners, are divided into three groups according to when they received their doctorates and/or entered the profession: before 1890, between 1890 and 1914, and after 1914. Of the fifteen individuals in the first group, one (T. W. Richards) held an American doctorate, seven held foreign ones (all of them from German universities) and another seven held none at all. The 50–50 split between Ph.D.s and non-Ph.D.s is also extant in Kevles's much larger group of productive physicists, chemists, and mathematicians, 1870 to 1890.[22] The same congruence is found to apply to those graduating between 1890 and 1914. By this time, American universities were much better equipped for graduate education; ten candidates thus held domestic Ph.D.s; three, foreign ones; and three, none at all. After 1914, all the candidates had earned their Ph.D.s from American universities.

The winners among the candidates have in common with Zuckerman's larger population of laureates that they received their doctorates at an early age and thus had a head start in their careers. The median age 26 years compares well with that of 25 for Zuckerman's laureates. By contrast, the nonwinners were a median 28 years old when they received their degrees, which is the same as Zuckerman's larger population of members of the National Academy of Sciences, 1900 to 1967. According to Zuckerman, "the run of doctorates in science in 1957" was a median of 30 years.[23] These differences cannot be explained solely by a selective educational process or by career strategies, as Zuckerman seems to think,

[22] Daniel J. Kevles and Carolyn Harding, *The physics, mathematics and chemical communities in America, 1870–1915: A statistical survey*, California Institute of Technology, Social science working paper no. 136 (Pasadena, 1977), Tables 5–12.

[23] Zuckerman, *Scientific elite*, p. 89.

but must also contain an element of native ability, which she does not discuss.

Master-apprentice relations

An important aspect of the training and socialization of the ultra-elite in Zuckerman's analysis is the apprenticeships served by future laureates under scientists who had already won the prize. This is also the mechanism whereby the ultra-elite becomes self-perpetuating. Of the 92 American laureates, she found that 48, or slightly more than half, had been trained by actual or future laureates. This figure is not far off the mark for laureates of every nationality, 41 percent of whom "have had at least one laureate master or senior collaborator." The percentage for the Americans would have been even higher, she claimed, had it not been for some solitary masters such as P. W. Bridgman, who restricted the number of his thesis students and thus had "only" one future laureate, as an apprentice. According to Zuckerman, this was John Bardeen, who won the prize twice (Ph 1956, 1972). (Bardeen's thesis director, as a point of fact, was not Bridgman but Eugene Wigner, who won the physics prize in 1963. Bardeen did not come into contact with Bridgman until he became a junior fellow at Harvard in 1935.[24])

This pairing off between Nobel masters and apprentices was not only characteristic of the postwar world of "big science" but could be extended back, Zuckerman claimed, through several generations of scientists. One such chain involved five generations of physicists and chemists: It started with Wilhelm Ostwald (Ch 1909) and ran through Walther Nernst (Ch 1920), Robert Millikan (Ph 1923), Carl Anderson (Ph 1936), to end (as of 1977) with Donald Glaser (Ph 1950).[25]

My comparison between the winning and the nonwinning candidates reveals that the winners were only slightly more likely than the nonwinners to have had a Nobel laureate master, when the latter is defined as the thesis adviser. True, some physics winners – C. J. Davisson, A. H. Compton, and C. D. Anderson – did their theses under the guidance of Nobel masters (O. W. Richardson in the case of the two former and R. A. Millikan for the latter), but so did the nonwinner

---

[24] Ibid., pp. 96–143; John Bardeen, "Reminiscences of early days in solid state physics," *Proceedings of the Royal Society of London* A371 (1980): 77–83.
[25] Zuckerman, *Scientific elite*, pp. 99–106.

I. S. Sprague (R. A. Millikan). Among the chemists, Langmuir was the only winner with a future laureate (Walther Nernst) as a thesis adviser; the others were the nonwinners G. N. Lewis (T. W. Richards) and A. A. Noyes (W. Ostwald). One is also struck by the number of laureates whose masters were not very prominent: Giauque, for instance, did his thesis at Berkeley under G. E. Gibson rather than G. N. Lewis; T. W. Richards at Harvard under J. P. Cooke rather than J. W. Gibbs, and E. O. Lawrence followed his master, W. F. G. Swann, a specialist in cosmic ray studies, from the University of Minnesota to Chicago and then to Yale, where he received his doctorate in 1925.[26]

Zuckerman's definition of a "master" covered not only thesis advisers but also those sought out for postdoctoral study. This provides a few more Nobel masters for the winners among the candidates in the period up to 1938: Richards, who spent a postdoctoral semester with Ostwald in Leipzig and Nernst in Göttingen (1895); and Urey, who did postgraduate work (1924–1925) in Bohr's institute in Copenhagen. This extended definition also has the effect, however, of increasing the number of nonwinning candidates apprenticed to Nobel masters. G. N. Lewis, for instance, excelled by having three Nobel masters: Richards, his thesis adviser, and Ostwald and Nernst with whom he spent a postdoctoral year (1901).

To fit so many of the laureates on to the procrustean bed of master-apprentice relations, Zuckerman had to go beyond her stated definition of an apprentice as a graduate student or postdoctorate and assume that a mere coincidence in time and place constituted a master-apprentice relation. One example will suffice here: A. A. Michelson, who is shown as having been apprenticed to Gabriel Lippmann (Ph 1908). During the winter of 1881–1882 that Michelson spent in Paris, however, he probably saw more of other French physicists specializing in optics – Alfred Cornu and Eleuthère Mascart, in particular – but of course they did not qualify as Nobel masters. Furthermore, Michelson was hardly an apprentice at this stage, his reputation being in fact so well established that when he appeared in Paris the French physicists thought he was the son of the "famous Michelson."[27]

---

[26] Heilbron and Seidel, *Lawrence and his laboratory*, p. 21.
[27] Dorothy Livingston Michelson, *The master of light: A biography of A. A. Michelson* (New York, 1973), pp. 86–88.

First jobs and subsequent employment

The candidates' training at elite institutions carried over to their first jobs. Slightly more than half of the candidates received their first appointment at an elite institution; for another half again this was also the institution that had granted them their doctoral degree. In the case of the laureates studied by Zuckerman, the corresponding figures are 65 and 40 percent. By the time the future laureates had been appointed to full professors, however, 78 percent had their appointments in elite institutions. By contrast, only 50 percent of the nonwinning candidates came to hold such appointments. The stronger concentration of Zuckerman's laureates at elite institutions reflects their having come of age professionally in the 1930s and 1940s, when American research universities had matured into institutions that could not only educate and train scientists but also sustain their careers.[28]

This was not so during the earlier period, when many of the nonwinning candidates embarked on their careers. Instead, they show a variety of first-job experiences: G. E. Hale, who on having received his B.A. from MIT, went back to his hometown of Chicago to found the Yerkes Observatory and the *Astrophysical Journal;* Robert W. Wood, who started out teaching at the University of Wisconsin before he was called to succeed Henry Rowland at Johns Hopkins University; or William D. Harkins, who taught chemistry at the University of Montana while preparing for his doctorate at Stanford University (1907). Alongside these, there were those who, like Zuckerman's laureates, started out as instructors at the universities where they had received their Ph.D.s and then stayed to become full professors: Moses Gomberg at Michigan, Theodore Lyman and T. W. Richards at Harvard, and M. I. Pupin at Columbia.

For those who had earned their doctorates after World War I, the fellowships of the National Research Council (NRC) became the royal route to permanent positions at elite institutions. Funded by the Rockefeller Foundation as part of a scheme to do for physics, chemistry, and mathematics what the Rockefeller Institute had done for medicine, the NRC awarded almost a thousand fellowships in those disciplines from the time the program got into full swing in the early 1920s until it was

---

[28] Roger L. Geiger, *To advance knowledge,* pp. 223–226.

discontinued in 1940. The large majority of the fellowships went to elite universities, in particular the California Institute of Technology and the University of California at Berkeley. In 1927, A. H. Compton, one of the first fellows, won the Nobel prize in physics. Among other American physics laureates before World War II, both C. D. Anderson and E. O. Lawrence, at Caltech and Berkeley, respectively, had been fellows.[29]

Whereas only 6 out of 92, or 7 percent, of the laureates studied by Zuckerman were employed by other than academic institutions, usually government or industry, this was the case for 10 out of 40, or 25 percent, of the candidates. Two prizes were awarded nonacademics, one to C. J. Davisson of Bell Telephone Laboratories and the other to Irving Langmuir of the research department of the General Electric Company. The 10 nonacademics among the candidates were evenly split between those working in government research offices, principally the U.S. Geological Survey and the U.S. Bureau of Standards, and in industrial research laboratories, such as Bell Telephone or General Electric. The former tended to be those who had received their doctorates or entered the profession before 1914 and the latter during the interwar period.

Honorific awards

According to Zuckerman, the superior performance of laureates throughout their careers – they publish early and copiously, they carry out their Nobel prizewinning research when in their late thirties or early forties, they continue to be more productive than other scientists of their age after they have won the prize – is confirmed by the rewards they receive in the form of promotions, positions of authority, and honorific awards. Honorific awards have been singled out here because they are less influenced by the vagaries of institutional praxis and academic politics than promotions. How, then, do the laureates and the nonwinning candidates compare with respect to the three major forms of honors studied by Cole and Zuckerman: prizes and medals, honorary doctorates, and election to domestic and foreign academies of sciences?

Scientific achievement was, and probably still is, recognized first by election to one's own national academy of science, in the United States,

[29] Tobey, *American ideology*, pp. 54–55; Kevles, *The physicists*, pp. 198–201, 219–220, 250; Heilbron and Seidel, *Lawrence and his laboratory*, pp. 13–14; and National Research Council, *National research fellowships, 1919–1938* (Washington, D.C., 1938), pp. 13–22.

Critical and empirical studies

the National Academy of Sciences (NAS). Foreign membership in one of the great academies of science – the Royal Society of London, the Académie des Sciences de Paris, the Prussian Academy of Sciences in Berlin – usually came later, if at all. With one exception, Fernando Sanford of Stanford University, the laureates and the nonwinning candidates were all members of the NAS. Furthermore, as Zuckerman pointed out, in the case of the laureates, election to the NAS almost always preceded the award of the Nobel prize.

The candidates' elections to foreign academies confirm the supposition that national honors usually precede international ones. Here, again, both winners and nonwinners were honored; of the nine candidates who were members of the Royal Society of London, the highest ranking foreign scientific society in the opinion of the physicists responding to the Coles' survey, three were laureates and six were not.[30] Among both groups membership in foreign academies was much more frequent before World War I caused irremediable damage to international exchanges of honors. Thus membership in foreign academies was concentrated in a handful of individuals who represented the first generation of candidates: G. E. Hale, who held eight such memberships; Henry Rowland, with the same number; and T. W. Richards and A. A. Michelson, with seven memberships each.

In the areas of prizes and medals and, to a lesser degree, honorary doctorates, the winners distinguished themselves from the nonwinners by collecting awards earlier in their careers and particularly by the international character of the awards. The only awards that count for the laureates are of course those received before the Nobel prize, as after the prize they were generally showered with honors.[31] Whereas the large majority of the chemists held the prestigious J. Willard Gibbs Medal of the American Chemical Society, for instance, the fact that the future laureates received this award before the Nobel prize, that is, generally before the age of 50, is another indication of their being more precocious than their colleagues. The same applies, although to a lesser degree, to both the Cresson Medal of the Franklin Institute and the Comstock Prize and Henry Draper Medal of the NAS. At the time they were awarded the Nobel prize, many of the future laureates had also received

---

[30] The members of the Royal Society were J. J. Abel, W. W. Campbell, W. Gibbs, G. E. Hale, I. Langmuir, A. A. Michelson, T. W. Richards, H. A. Rowland, and R. W. Wood.
[31] Zuckerman, *Scientific elite*, pp. 236–238.

the more prestigious honorary doctorates – from Harvard, Princeton, Columbia, Yale – which was not true for the nonwinning candidates at the same age.

The winners and the nonwinners differed most significantly when it came to international awards, mainly the medals that the Royal Society of London bestowed on prominent foreign scientists: the Copley, Rumford, Davy, and Hughes medals. The majority of the winners and the two most prominent runners-up, G. E. Hale and G. N. Lewis, all received one or more of these honors in the years of active candidacy for the Nobel prize. A. A. Michelson, for instance, received the Copley in 1907, only a month before the announcement of his Nobel award; Hale also received the Copley but much later in life; C. J. Davisson, I. Langmuir, E. O. Lawrence, R. A. Millikan were all honored with the Hughes Medal before their Nobel prize; and G. N. Lewis and T. W. Richards with the Davy both while they were being considered for the prize. In this they did not differ from the overall group of prizewinners in the period before World War I. That the prizewinners were most often those who had received these other international honors has been taken as an indication of how the Swedish prize juries, when they were building up the institution, took their cues from these already existing awards, whose longer history and prestige could "rub off" on the new prizes.[32]

From nominal to functional elites

In her study, Zuckerman chose to delineate her ultra-elite of American scientists nominally, that is, she included everybody whom the prize juries in Sweden had designated a winner of the Nobel prizes in physics, chemistry, and physiology or medicine. In an analogous manner, one could designate Supreme Court justices as the ultra-elite of judges or the chief executive officers of the Fortune Five Hundred as the ultra-elite of business. Because they can be named and enumerated does not resolve the more basic difficulty of separating the ultra-elite from the elite.[33]

---

[32] Crawford, *The beginnings of the Nobel institution*, pp. 202–203.
[33] The sociological literature tracing the social origins of different categories of elites up until the early 1960s is surveyed in Suzanne Keller, *Beyond the ruling class: Strategic elites in modern society* (New York, 1963), pp. 292–326. There does not seem to be any analogous work for more recent

Critical and empirical studies

This quandary was demonstrated by the comparison between the winners and the nonwinning candidates for the Nobel prizes whose similarities with respect to the attributes studied were in fact greater than their differences. The two attributes where the winners and the nonwinning candidates differed make it clear that the composition of Zuckerman's ultra-elite of American laureates largely depended on successive prize decisions by scientific corporations in Sweden. One was the dearth of industrial scientists among the laureates as compared to the nonwinning candidates. As has already been pointed out, this was a consequence of the prize juries showing only limited interest in awards in the area of technology during the period of Zuckerman's study. The other was the stronger international bent of the winners as compared to the nonwinners. Although observed here only with respect to honorific awards, it would not be surprising if the winners were also found to be more international than local in their publication and citation patterns. Here, again, one sees the hand of the prize juries, specifically their concern to remain in the mainstream of international specialty orientations in physics and chemistry.

It is precisely this concern that Zuckerman failed to appreciate, as she did not view the laureates' work historically in relation to developments of theory and method in the three disciplines she treated. She overlooked that her scientific elite covers almost seven decades of intellectual and institutional development of American science, and hence cannot be treated as a monolith. If viewed with an eye toward history, the basic research orientation in theoretical microphysics and biochemistry that had become the hallmark of the laureate population in the late 1970s would have appeared, in her analysis, not as immanent features of the American scientific enterprise but rather as the high point of a particular historical development.

What, then, about the elite status of the overall group of candidates? Is it possible to demarcate this group in relation to an "extended elite," however defined, and to delineate that "extended elite" with respect to the mass of scientists that surrounds it? If the elite attributes surveyed

times. The most original work on elites was done at the turn of the century by German and Italian sociologists and social theorists (Karl Mannheim, Max Weber, Georg Simmel, Vilfredo Pareto, and Gaetano Mosca, among others). For a guide to the German writings, see Walther Struwe, *Elites against democracy: Leadership ideals in bourgeois political thought in Germany, 1890–1933* (Princeton, N. J., 1973).

were used, such an exercise would probably yield a gradual fading of the elite into the nonelite rather than a sharp cutoff between the two. The concept is not very interesting, however, when used to demonstrate individually that some scientists are better educated, work in better endowed institutions, and garner more honors than others. To become fruitful, inquiries into elites have to ask questions about some of the functions that these have performed, or can perform, collectively. Such inquiries would probably point up not a single elite or ultra-elite, but different kinds. In the following discussion, functions that were important when it came to organizing and running the national scientific enterprise are sketched.

Institutional elitism

For elite institutions to emerge, there clearly had to be individuals and groups capable of promoting them. The trend in America at the end of the nineteenth century, in contrast to Germany, for instance, was to make these science-based. This required raising the funds for building and equipping laboratories, recruiting faculty, and reforming curricula, all tasks undertaken by entrepreneurial university presidents and department chairs starting in the 1870s and continuing to the present day. It was this kind of best-science elite and elitism that Henry Rowland advocated in his 1883 address to the American Association for the Advancement of Science when he called for the concentration of resources in monies and personnel to a few first-class universities.[34]

The realization of Rowland's program is exemplified by the triumvirate G. E. Hale, Robert Millikan, and A. A. Noyes, who joined forces in the early 1920s to turn the Throop College of Technology in Pasadena into a major research university, the California Institute of Technology. There Hale, who had had his base at the Mount Wilson Observatory since early in the century, enrolled Millikan and Noyes – from Chicago and MIT, respectively – in an ambitious program that drew on private and public sources of financing to advance basic science.[35] All three men were candidates for the Nobel prizes, but only Millikan was a winner. Although solid scientific achievements were certainly important

---

[34] Kevles, *The physicists*, pp. 43–46.
[35] Geiger, *To advance knowledge*, pp. 183–191.

Critical and empirical studies

in acceding to the institutional elite, these did not have to be of Nobel caliber. This is shown by the majority of those who built and operated the physics and chemistry departments at elite institutions. Although the "stars" were certainly important, the effectiveness of institutional elites probably depended as much, or more, on local as on national or international visibility and commitments.

Political elitism

As defined by Kevles, political elitism resulted from the assumption of many scientists that "they were a select group who deserved to exercise power with only limited accountability."[36] This principle, first put to work for science and scientists in the federal government in the 1880s by John Wesley Powell, head of the Geological Survey, was to dominate all future relations of science and government. It would give basic science a home in countless civilian federal agencies; it would turn the National Academy of Sciences and its executive arm, the National Research Council, into power brokers between the government and the scientific community; and it would permit the massive mobilization of scientists in World War I and, perhaps more important, their demobilization and return to academic life. Mapping the political elite of scientists in their interrelations with the federal government up to World War II is clearly a massive task, but feasible and useful. For example, it could give perspective to the political elitist roles of particularly ubiquitous scientists, such as Hale and Millikan, roles that may have been obscured by their emergence as public figures.

Conclusions

The high visibility of Nobel prizewinners on the American scientific scene makes them appear as the select few, predestined to membership in an ultra-elite. In Zuckerman's opinion, this ultra-elite reflects immanent features – elitism, stratification, and the accumulation of advantages through the reward system of science – that have determined the organizational structure of American science in the twentieth century and

[36] Henry Augustus Rowland, "A plea for pure science," in *The physical papers of Henry Augustus Rowland* (Baltimore, 1902), pp. 593–613; and Kevles, *The physicists*, pp. 43–44.

that, furthermore, promote progress in science. The comparison between winning and nonwinning candidates in physics and chemistry during the period 1901 to 1939 carried out in this study has shown that the similarities between the two groups with respect to the characteristics that Zuckerman identified as elite attributes were larger than the differences. The two primary differences – the dearth of industrial scientists among the winners and their higher international profile – can be explained by the ultra-elite of laureates having been created through successive decisions by the Swedish Nobel awarders. This points up the fundamental ambiguity of Zuckerman's use of the prize adjudications by scientific corporations in Sweden to delineate a national ultra-elite in the United States.

The notion of scientific elite would have more meaning if, rather than attempt to separate the scientific elite from the nonelite, one explored the functions performed by different elites. What is termed "institutional" and "political" elitism could give insight into the creation of elite universities and their continued domination over American science and the conditions of scientists' participation in government research activities. Even in this extended sense, however, the notion does not provide much specific insight into how different kinds of elites and elitism advance scientific knowledge. This necessarily impairs its explanatory value in the history and sociology of science.

Bibliographical essay

The literature about nationalism is voluminous, especially for the period 1880 to 1939, but very little bears on the sciences. This applies to the "founding fathers" of the academic study of the phenomenon – Hans Krohn, *The idea of nationalism: A study in its origins and background* (New York, 1960); and Carleton B. Hayes, *The historical evolution of modern nationalism* (New York, 1931) – as well as to such works as E. J. Hobsbawm, *Nations and nationalism since 1789: Programme, myth, reality* (Cambridge, 1990); Michael Hughes, *Nationalism and society: Germany, 1800–1945* (London, 1988); and Boyd C. Shafer, *Faces of nationalism: New realities and old myths* (New York, 1972). An exception is Ernst Gellner, *Nations and nationalism* (Ithaca and London, 1983), whose model of nationalism I have used in this book (Chapter 2). All of these works contain bibliographies; the one in Hobsbawm is particularly useful. A survey of common usage of the terms nation and nationalism in different languages is found in *Nationalism. A report by a study group of members of the Royal Institute of International Affairs* (New York, 1966).

The literature concerning chauvinism among scientists during and after World War I is found in the notes to Chapter 3. Scientific rivalry prior to the war is described in Harry W. Paul, *The sorcerer's apprentice: The French scientist's image of German science, 1840–1919* (Gainesville, 1972). The literature concerning science linked to imperialism and colonialism is growing rapidly and cannot be surveyed exhaustively here. Among the more important titles two by Lewis Pyenson should be noted: *Cultural imperialism and exact sciences: German expansion overseas, 1900–1930* (New York, 1985) and *Empire of reason: Exact sciences in Indonesia, 1840–1940* (Leyden, 1989). See also the collective volume, Nathan

Bibliographical essay

Reingold and Marc Rothenberg, eds., *Scientific colonialism: A cross-cultural comparison* (Washington, D.C., 1987).

The history of internationalism has been the object of far less academic work than nationalism, this gap partly being filled by the massive study of F. S. L. Lyons, *Internationalism in Europe, 1815–1914* (Leyden, 1963). It contains a useful section on international scientific collaboration (Part II, d). Overviews on internationalism in science are also found in Elisabeth Crawford, "The universe of international science, 1880–1939," in Tore Frängsmyr, ed., *Solomon's house revisited: The organization and institutionalization of science.* Proceedings of Nobel Symposium 75 (Canton, Mass., 1990), pp. 251–269; and Brigitte Schroeder-Gudehus, "Nationalism and internationalism," in G. N. Cantor et al., eds., *Companion to the history of modern science* (London and New York, 1989), pp. 909–919. International scientific relations in the interwar period are treated in two books by Brigitte Schroeder-Gudehus, *Deutsche Wissenschaft and internationale Zusammenarbeit, 1914–1928* (Geneva, 1966), and *Les scientifiques et la paix: La communauté scientifique internationale au cours des années 20* (Montreal, 1978).

As one of the major international scientific institutions of the early twentieth century, the Nobel institution gave entry into much international activity at that time. Works illuminating different aspects of the international working of the institution are Elisabeth Crawford, *The beginnings of the Nobel institution: The science prizes 1901–1915* (Cambridge and Paris, 1984); the papers presented at the roundtable held at Nobel Symposium 52 (1981) and published in Carl Gustaf Bernhard, Elisabeth Crawford, and Per Sörbom, eds., *Science, technology and society in the time of Alfred Nobel* (Oxford, 1982); and Elisabeth Crawford, J. L. Heilbron, and Rebecca Ullrich, *The Nobel population, 1901–1937: A census of the nominators and nominees for the prizes in physics and chemistry* (Berkeley and Uppsala, 1987), the basis of the four studies presented in this book.

There does not exist a general social history of science that would provide background for the problems treated in the four studies of the Nobel population. The work that comes closest to doing so is Joseph Ben-David, *The scientist's role in society: A comparative study* (Englewood Cliffs, N.J., 1971).

Center-periphery considerations applied to the world economy are set out in Fernand Braudel, *Civilisation matérielle, économie et capitalisme, XV–XVIII siècle*, vol. 3, *Le temps du monde* (Paris, 1979), pp. 12–33. Other

Bibliographical essay

works dealing with center-periphery relations are by Edward Shils, *Center and periphery: Essays in macrosociology* (Chicago and London, 1975), and *The intellectuals and the powers and other essays* (Chicago and London, 1972), pp. 355–371. The specific problems of center-periphery relations in science are discussed in Rainald von Gizycki, "Center and periphery in the international scientific community: Germany, France and Great Britain in the 19th century," *Minerva* 11 (1973): 474–494; and Gabor Pallo, "Some conceptual problems of the center-periphery relationship in the history of science," *Philosophy and Social Action*, 13 (1987): 27–32.

A problem that Ben-David does not address is the elite conception of science that runs through the four studies and is discussed explicitly in Chapter 6. The elite view of the social organization of science is set forth in Jonathan R. Cole and Stephen Cole, *Social stratification in science* (Chicago, 1973); and Harriet Zuckerman, *Scientific elite: Nobel laureates in the United States* (New York, 1977). More general works dealing with elites and their historical roles in society are Suzanne Keller, *Beyond the ruling class: Strategic elites in modern society* (New York, 1963); and Walther Struwe, *Elites against democracy: Leadership ideals in bourgeois political thought in Germany, 1890–1933* (Princeton, N.J., 1973).

Finally, there is a large literature concerning the national scientific enterprises brought to the fore in the four studies. For Germany, the disciplines (physics and chemistry) featured in the studies receive extensive treatment in Christa Jungnickel and Russell McCormmach, *Intellectual mastery of nature. Theoretical physics from Ohm to Einstein* (2 vols.), vol. 1, *The torch of mathematics, 1800–1870*, and vol. 2, *The now mighty theoretical physicists, 1870–1925* (Chicago, 1986); and Jeffrey A. Johnson, *The kaiser's chemists: Science and modernization in imperial Germany* (Chapel Hill, 1990). The work by Jungnickel and McCormmach includes German-speaking areas outside the Reich. More detailed information about Austrian physics is found in Berta Karlik and Erich Schmid, *Franz Serafin Exner und sein Kreis: Ein Beitrag zur Geschichte der Physik in Österreich* (Vienna, 1982). Social and political developments are reflected in the histories of the two major scientific institutions, founded during the Wilhelmian era. These works are David Cahan, *An institute for an empire: The Physikalisch-Technische Reichsanstalt, 1871–1918* (Cambridge, 1989); and Rudolf Vierhaus and Bernhard vom Brocke, eds., *Forschung im Spannungsfeld von Politik und Gesellschaft:*

Bibliographical essay

Geschichte und Struktur der Kaiser-Wilhelm-/Max-Planck-Gesellschaft (Stuttgart, 1990). The question of Germany hegemony in science is discussed more generally in Alan Beyerchen, "On the stimulation of excellence in Wilhelmian science," in Jack R. Dukes and Joachim Remak, eds., *Another Germany: A reconsideration of the imperial era* (Boulder, Colo., and London, 1988), pp. 139–168. J. L. Heilbron, *The dilemmas of an upright man: Max Planck as spokesman for German science* (Berkeley, 1986), illuminates the rise and fall of German science.

For the United States, an overview is provided in Alexandra Oleson and John Voss, eds., *The organization of knowledge in modern America* (Baltimore, 1979). The main disciplinary history is Daniel J. Kevles, *The physicists: The history of a scientific community in modern America* (New York, 1978). Some of the materials for a history of chemistry are found in Arnold Thackray, Jeffrey L. Sturchio, P. Thomas Carroll, and Robert Bud, *Chemistry in America, 1876–1976: Historical indicators* (Dordrecht, 1985). The organization and politics of the enterprise are discussed in Roger L. Geiger, *To advance knowledge: The growth of American research universities, 1900–1940* (Oxford, 1986); and Ronald C. Tobey, *The American ideology of national science 1919–1930* (Pittsburgh, 1971).

For France, Harry W. Paul, *From knowledge to power: The rise of the science empire in France, 1860–1939* (Cambridge, 1985), provides a comprehensive treatment of the enterprise. More specialized aspects are discussed in Robert Fox and George Weisz, eds., *The organization of science and technology in France, 1808–1914* (Cambridge, 1980); Mary Jo Nye, *Science in the provinces: Scientific communities and provincial leadership in France, 1860–1930* (Berkeley, 1986); and Dominique Pestre, *Physique et physiciens en France, 1918–1940* (Paris and Montreux, 1984).

Index

~~~~~~

(t = table; f = footnote)

Index

Index

Index

Index

Index

Index

Printed in the United States
By Bookmasters